Allah and Space

Allah and Space

Dr Anwar El-Shahawy

Professor of physical Chemistry

To order additional copies of this book, contact:
Xlibris Corporation
0-800-644-6988
www.xlibrispublishing.co.uk
Orders@xlibrispublishing.co.uk
301030

Contents

2—Holocaust

Preface

The great importance of the Space in our life gives us an attention to think more about its interference in our life. Space is the arenas in which all physical events take place. An event is in a point of Space specified by its place and in a specific time. For example, the motion of a planet around a star may be described in a particular type of Space in certain time, or the motion of light around a revolving star may be described in another type of Spacetime continuum. An event is in a unique position at a unique time. However, physical event occurs at certain moment of time in a given region of Space. This attracts our attention about time. Is the time created by the energy or energy is created via the time according to Newtonian mechanics? What is the importance of the Spacetime continuum with respect to the life of the humanbeings? What is the relationship of this Spacetime continuum with the God of the Universe? From the importance of the Space and the time in our life we must think more about the Spacetime problem with respect to our Universe. What is the relationship of this problem with the religions? Try to find out the answers through this book. This text has exposed to some future events with respect to the Holy Quran such as Invasion of the Space, Holocaust and some astronomical events. Enjoy with the reading of this text.

The Author
Dr Anwar El-Shahawy
Professor of Physical Chemistry

The Space

When we are dealing with the space we aim to empty space. Where is the space? We know that atom which is the unit of any material is consisted of a very tiny heavy nucleus bearing positive charge and electrons of negative charge revolving around the nucleus in the center. But what is the ratio between the volume of the nucleus and the atom? it is 10^{-12} i.e. if you imagine a nucleus of $1cm^3$ volume therefore it should be in the center of an imaginary atom of 1 million cubic meter. This means that the big component of all atoms of anything is the space filling the atom and in the same time surrounds it. Hence, the space occupies all the molecules and around them. Therefore this space is the base of the existence of the world. On the other hand, space penetrates the material and surrounding it. Hence space is always bigger than the material whatever it is big or small. When you look around yourself to see something around you but you do not ask yourself what is between your eyes and this matter? May be nothing. But when you think more and more you will say that is air between your eyes and all things you see. But, when you think more and more you find out that air molecules, N_2, O_2 and CO_2, nitrogen, oxygen and carbon dioxide molecules are in between your eyes and all things around you. But, when you think deeply, you will ask yourself where those molecules are? Of course, the answer will jump to your mind telling you that all these molecules of air move inside the space. Therefore space is inside the matter and surrounding it. Extension of this idea, you will find out that space is inside yourself and surrounding yourself. But the space inside yourself contains atoms, molecules, tissues, organs and finally your body as a whole but the space around yourself contains the molecules of air. So when you see something around yourself you see them through the space. So space is very transparent to permit the vision through. But may be a question jumps to your mind; is the space being the same every where? Physicists Michelson and Moorly 1887 found out by their famous experiment that the speed of light in all directions is the same which is always equal to a universal constant value 3×10^{10} cm s^{-1} or 3×10^5 km s^{-1}. This means that space is the same every where. Also the light coming from far stars has the same velocity 3×10^5 km s^{-1} as well as the light coming from the sun star. This gives us an impression that space is the same from the earth planet to the heaven or to the seventh sky which have resemblance with the seven orbitals of the atom [May be].

When you breathe air (inspiration and expiration) you withdraw molecules of the air from the space into the space of your lungs and pushing air molecules from the space of your lungs to the big space surrounding yourself. When you separate the space of your lung with the big space surrounding you by closing the holes of your nose you will die. This means simply the life of any animal is the communication between the space of the lung with the big space surrounding the animal as well as the human beings or every thing.

When you hear a voice from another person, you receive the sound waves, the compressed and evacuated waves of air molecules, to enter through the space in your external auditory meatus in your ear to arrive to tympanic membrane inside the ear to hear this voice. So you hear the voice not only by your ear but also by the aid of the space in which the air molecules move in compressive and evacuated waves.

When you see television, you will find out that this device does not work by itself only but this device receives the electromagnetic waves from the satellite through the space to see the television show. So, the television device works by the nature of the space as well as when you talk in your mobile; of course your device receives the microwaves through the space to transmit the voice. The final conclusion which jumps to your mind is that you see through the space and hear by the aid of the space. To understand how is the space surrounding us? Imagine a glass sinks in water, you will find that water is inside the glass and surrounding it. So you will find out that when you move on the land in front of you, you penetrate this space and the air molecules replace your behind to keep the pressure around you being constant to keep pressure being equal always one atmosphere. Also, when you move to your back, you penetrate the space and the air molecules replace yourself in front of you, finally you will find out that you must have the space always in front of you and behind you to live in the earth planet, to do your work to do any thing you need. Finally your motion on the land is the replacement of the air molecules in the space wherever your direction.

Also, even when you eat, you take food from the space in front of you to the space of your mouth to chew the food and then to swallow it to the space of your stomach. Oh . . . . You will find out that space is the big partner with you every instant from your birthday to your death day as well as your parents and all people in the past and future.

Really did you think like this? So you will find out that space is the mysterious of the existence. More over, this space can contain everything whatever it is but is not affected by any thing and retains his properties after leaving the matter from its position to retain its properties without any effect. Also we must remember that the importance of the surrounding of the space to the matter is preservation the character of the matter from decay.

Do you remember the explosion of atomic bombs in Japan 1945, Nagasaki and Hiroshima? Every thing had been damaged and destroyed strongly but the space was not affected absolutely. Do you remember that the space is inside the sun star and surrounding it although the temperature inside the sun maintains several millions of temperature degrees, so the space is very strong or the strongest over the entire universe! So we can abbreviate some of the properties of this space as follows:

1. Space is the widest i.e. bigger than any thing, electron, proton, atom, molecule, animals, plants, planets, stars and galaxies.
2. Space surrounds every thing and anything and can be described as circumambient with all the matters.
3. Space is very transparent or very diaphanous from which we can see every things through the reflected light, i.e. fortune-teller.
4. Space is earshot as we mentioned before.
5. Space is bigger than anything.
6. Space is softness or tenderness. All the matters can move in the space easily without remarkable resistance [Affable].

7.   Space is superintendent for the events or the motion of the materials [Watcher].

8.   Space has the ability to contain the matter.

9.   Space has the ability to contain the magnetic field.

10. Space has the ability to contain the electric field.

11. Space has the ability to contain the electromagnetic waves (light of different wavelengths).

12. Space is the strongest over the entire universe since it contains the stars and planets in their huge galaxies. Any star can be considered as a nuclear reactor in which the nuclear fusion reactions take place to release a huge amount of heat to achieve several millions of temperature degrees. In this occasion, the temperature of the surface of the sun star is about 6000°C.

13. The circumambient property of the space around the materials conserves them from the decay.

14. Space is before the matter or space is the first. [Space is Eternal.]

15. Space isn't affected by the matter But Space define the shape of the matters.

16. Space penetrates the matter surrounding its components (Atoms or Molecules) inside the matter. Although the Space inside the matter it is not affected by the matter constituents (Atoms or Molecules). Therefore Space is surrounding the matter externally and inside the mater without any effect on the space.

17. In the commencement of the universe, there was a huge amount of hot cosmic gas as very dark fumes in its plasma state in the space; the universe expanded from the extremely dense and very dark fume of cosmic gas in which nuclear fusion reactions of hydrogen nuclei took place to produce a huge amount of all the elements of the periodic table and continues to expand till today as stars and planets in their galaxies aggregations in the space till now. A common analogy explains that space itself is expanding, carrying galaxies with it.

18. Definition of the matter is that occupies definite part of the space.

19. Space is the widest over all the matters from atoms to galaxies. From the start of big bang the very hot condensed materials expand in the Space since 14 billion years and going to expand till the gravitational attractive force between the galaxies overcomes

the spreading energy then the galaxies containing solar systems attract each other to be crashed strongly to form the atomic particles to form the dark fumes as the beginning before the big bang. which is the commencement of the creation of the universe.

ء

The unity of the space

The arrival of sun light to the earth planet proves that the space from the sun to the earth is the same and the arrival of the light of the stars with same light velocity 3×10^5 km s^{-1} also proves that the space from the stars to the earth planet is the same. The motion of the stars and the planets throughout the space proves that the space is the same overall the universe. The motion of animals on the surface of the earth planet proves the unity of the space since the matter moves overall the universe in the space. So the space is the same with respect to the matter as well as to the electromagnetic waves (light) since light has its corpuscular property. This means that space is the same for anything.

Did you think about time? What is the time? Do you think that time is dependent on the space? But how? When you think that the earth planet revolves around the sun star one compete circle, it makes one year. So it is necessary for the earth to find space around the sun star to move around it to perform one year. So the earth planet revolves one circle around the sun and it performs a rotational energy unit for each circle around the sun. So every rotational energy unit is corresponding to one year. The repetition of this energy unit means the continuation of the sun-earth system in the space. So we can introduce a general definition of the time from a general point of view as follows:

Time is the resultant of the repetition of the energy unit in any system.

To understand this definition, we imagine the pulse as the energy unit in the human being; therefore the continuation of this human being depends on the repetition of this energy unit in a system or the pulse for human being. Therefore time is relative from any system to other, depending on the absolute value of the energy unit inside the system. So the time of the solar system is different from the time of the human being and the time of human being is different from the time of plant.

The time of galaxy is different from the time of the solar system. Do you think about the energy required to move the universe one day? To know this energy, we must know the energy unit in each system in one day of the earth planet in the all over the universe such as solar systems, all animals, all human beings, all the plants, all galaxies, evils, Jenny and angels then we can evaluate the energy required to move the universe one day forward. Did you imagine this huge amount of this energy even required to move the universe one second?

Continuation of the systems:

Since the time is the resultant of the repetition of the energy unit in systems, therefore the continuation of any system depends on this repetition. This leads us that all systems: solar system (planets revolve around stars), galaxies, animals, and other systems such as evils and angels, need a center of energy supplying them with the repetition of energy unit for each system to be continued in the life as it can be considered Time. This center of energy supplies all the systems by the energy units for their continuation. This can be expressed in another words; this center of the energies communicates with all the systems to repeat the energy unit in each system for continuation in the space as can be considered as creation of the time.

Time and Space

Time has been defined as the continuum in which events occur in succession from the past to the present and on to the future. Time has also been defined as a one-dimensional quantity used to sequence events, to quantify the durations of events and the intervals between them.

Time has also been defined as a one-dimensional quantity used to sequence events, to quantify the durations of events and the intervals between them, and (used together with other quantities such as space) to quantify and measure the motions of objects and other changes. Time is quantified in comparative terms (such as longer, shorter, faster, quicker, slower) or in numerical terms using units (such as seconds, minutes, hours, days). Time has been a major subject of religion, philosophy, and science, but defining it in a non-controversial manner applicable to all fields of study has consistently eluded the greatest scholars.

Among prominent philosophers, there are two distinct viewpoints on time. One view is that time is part of the fundamental structure of the universe, a dimension in which events occur in sequence. Sir Isaac Newton subscribed to this realist view, and hence it is sometimes referred to as Newtonian time. Time travel, in this view, becomes a possibility as other "times" persist like frames of a film strip, spread out across the time line. The opposing view is that *time* does not refer to any kind of "container" that events and objects "move through", nor to any entity that "flows", but that it is instead part of a fundamental intellectual structure (together with space and number) within which humans sequence and compare events. This second view, in the tradition of Gottfried Leibniz and Immanuel Kant, holds that *time* is neither an event nor a thing, and thus is not itself measurable nor can it be travelled.

Temporal measurement has occupied scientists and technologists, and was a prime motivation in navigation and astronomy. Periodic events and periodic motion have long served as standards for units of time. Examples include the apparent motion of the sun across the sky, the phases of the moon, the swing of a pendulum, and the beat of a heart of humanbeing. Currently, the international unit of time, the second, is defined in terms of radiation emitted by caesium atoms (see below). Time is also of significant social importance, having economic value ("time is money") as well as personal value, due to an awareness of the limited time in each day and in human life spans.

From Wikipedia, the free encyclopedia August 23, 2010

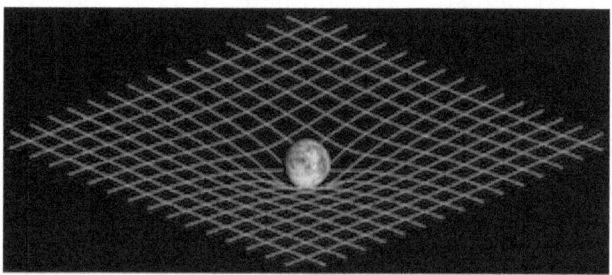

Two-dimensional analogy of space-time distortion. Matter changes the geometry of space-time, this (curved) geometry being interpreted as gravity. White lines do not represent the curvature of space but instead represent the coordinate system imposed on the curved space-time, which would be rectilinear in a flat space-time.

In physics, space-time (or *space-time; or space/time*) is any mathematical model that combines space and time into a single continuum. Space-time is usually interpreted with space being three-dimensional and time playing the role of a fourth dimension that is of a different sort from the spatial dimensions. According to certain Euclidean space perceptions, the universe has three dimensions of space and one dimension of time. By combining space and time into a single manifold, physicists have significantly simplified a large number of physical theories, as well as described in a more uniform way the workings of the universe at both the super-galactic and subatomic levels.

In classical mechanics, the use of Euclidean space instead of space-time is appropriate, as time is treated as universal and constant, being independent of the state of motion of an observer. In relativistic contexts, however, time cannot be separated from the three dimensions of space, because the observed rate at which time passes for an object depends on the object's velocity relative to the observer and also on the strength of intense gravitational fields, which can slow the passage of time.

Mathematical concept of space-time

The first reference to space-time as a mathematical concept was in 1754 by Jean le Rond d'Alembert in the article *Dimension* in Encyclopedia. Another early venture was by Joseph Louis Lagrange in his *Theory of Analytic Functions* (1797, 1813). He said, "One may view mechanics as geometry of four dimensions, and mechanical analysis as an extension of geometric analysis".

After discovering quaternion, William Rowan Hamilton commented, "Time is said to have only one dimension, and space to have three dimensions . . . . The mathematical quaternion partakes of both these elements; in technical language it may be said to be 'time plus space', or 'space plus time': and in this sense it has, or at least involves a reference to, four dimensions. And how the One of Time, of Space the Three, Might in the Chain of Symbols girdled be." Hamilton's bi-quaternion, which have algebraic properties sufficient to model space-time and its symmetry, were in play for more than a half-century before formal relativity. For instance, William Kingdon Clifford noted their relevance.

Another important antecedent to space-time was the work of Clerk Maxwell as he used partial differential equations to develop electrodynamics with the four parameters. Lorentz discovered some invariance of Maxwell's equations late in the 19th century which were to become the basis of Einstein's theory of special relativity. Fiction authors were also on the game as mentioned above. It has always been the case that time and space are measured using real numbers, and the suggestion that the dimensions of space and time are comparable could have been raised by the first people to have formalized physics, but ultimately, the contradictions between Maxwell's laws and

Galilean relativity had to come to a head with the realization of the import of finitude of the speed of light.

While space-time can be viewed as a consequence of Albert Einstein's 1905 theory of special relativity, it was first explicitly proposed mathematically by one of his teachers, the mathematician Hermann Minkowski, in a 1908 essay building on and extending Einstein's work. His concept of Minkowski space is the earliest treatment of space and time as two aspects of a unified whole, the essence of special relativity. The idea of Minkowski space also led to special relativity being viewed in a more geometrical way, this geometric viewpoint of space-time being important in general relativity too. (For an English translation of Minkowski's article, see Lorentz et al. 1952.) The 1926 thirteenth edition of the Encyclopedia Britannica included an article by Einstein titled "Space-Time".

Basic concepts

Space-times are the arenas in which all physical events take place—an event is a point in space-time specified by its time and place. For example, the motion of planets around the sun may be described in a particular type of space-time, or the motion of light around a rotating star may be described in another type of space-time. The basic elements of space-time are events. In any given space-time, an event is a unique position at a unique time. Because events are space-time points, an example of an event in classical relativistic physics is (x,y,z,t), the location of an elementary (point-like) particle at a particular time. A space-time itself can be viewed as the union of all events in the same way that a line is the union of all of its points, organized into a manifold (a locally flat metric space).

A space-time is independent of any observer. However, in describing physical phenomena (which occur at certain moments of time in a given region of space), each observer chooses a convenient metrical coordinate system. Events are specified by four real numbers in any such coordinate system. The trajectories of elementary (point-like) particles through space and time are thus a continuum of events called the world line of the particle. Extended or composite objects (consisting of many elementary particles) are thus a union of much worldliness twisted together by virtue of their interactions through space-time into a "world-braid" (permitting a fascinating connection with the myth of the Moirae to be made).

However, in physics, it is common to treat an extended object as a "particle" or "field" with its own unique (e.g. center of mass) position at any given time, so that the world line of a particle or light beam is the path that this particle or beam takes in the space-time and represents the history of the particle or beam. The world line of the orbit of the Earth (in such a description) is depicted in two spatial dimensions x and y (the plane of the Earth's orbit) and a time dimension orthogonal to x and y. The orbit of the Earth is an ellipse in space alone, but its world line is a helix in space-time.

Space-time in Relativity Theory

In general relativity, it is assumed that space-time is curved by the presence of matter (energy), this curvature being represented by the Riemann tensor. In special relativity, the Riemann tensor is identically zero, and so this concept of "non-curvedness" is sometimes expressed by the statement *Minkowski space-time is flat.*

Many space-time continua have physical interpretations which most physicists would consider bizarre or unsettling. For example, a compact space-time has closed, time-like curves, which violate our usual ideas of causality (that is, future events could affect past ones). For this reason, mathematical physicists usually consider only restricted subsets of all the possible space-times. One way to do this is to study "realistic" solutions of the equations of general relativity. Another way is to add some additional "physically reasonable" but still fairly general geometric restrictions and try to prove interesting things about the resulting space-times. The latter approach has led to some important results, most notably the Penrose-Hawking singularity theorems.

Quantized space-time

In general relativity, space-time is assumed to be smooth and continuous—and not just in the mathematical sense. In the theory of quantum mechanics, there is an inherent discreteness present in physics. In attempting to reconcile these two theories, it is sometimes postulated that space-time should be quantized at the very smallest scales. Current theory is focused on the nature of space-time at the Planck scale. Causal sets, loop quantum gravity, string theory, and black hole thermodynamics all predict a quantized space-time with agreement on the order of magnitude. Loop quantum gravity makes precise predictions about the geometry of space-time at the Planck scale.

Character of space-time

There are two kinds of dimensions, spatial (bidirectional) and temporal (unidirectional). Let the number of spatial dimensions be N and the number of temporal dimensions be T. That $N = 3$ and $T = 1$, setting aside the compactified dimensions invoked by string theory and undetectable to date, can be explained by appealing to the physical consequences of letting N differ from 3 and T differ from 1. The argument is often of an entropic character.

Immanuel Kant argued that 3-dimensional space was a consequence of the inverse square law of universal gravitation. While Kant's argument is historically important, John D. Barrow says that it " . . . gets the punch-line back to front: it is the three-dimensionality of space that explains why we see inverse-square force laws in Nature, not vice-versa." (Barrow 2002: 204). This is because the law of gravitation (or any other inverse-square law) follows from the concept of flux and the proportional relationship of flux density and the strength of field. If $N = 3$, then 3-dimensional solid objects have surface areas proportional to the square of their size in any selected spatial dimension. In particular, a sphere of radius r has area of $4\pi r^2$. More generally, in a space of N dimensions, the strength of the gravitational attraction between two bodies separated by a distance of r would be inversely proportional to r^{N-1}.

In 1920, Paul Ehrenfest showed that if we fix $T = 1$ and let $N > 3$, the orbit of a planet about its sun cannot remain stable. The same is true of a star's orbit around the center of its galaxy. Ehrenfest also showed that if N is even, then the different parts of a wave impulse will travel at different speeds. If $N > 3$ and odd, then wave impulses become distorted. Only when

$N = 3$ or 1 are both problems avoided. In 1922, Hermann Weyl showed that Maxwell's theory of electromagnetism works only when $N = 3$ and $T = 1$, writing that this fact " . . . not only leads to a deeper understanding of Maxwell's theory, but also of the fact that the world is four dimensional, which has hitherto always been accepted as merely 'accidental,' become intelligible through it."[12] Finally, Tangherlini showed in 1963 that when $N > 3$, electron orbital around nuclei cannot be stable; electrons would either fall into the nucleus or disperse.

Properties of Dimensional Space-times

Max Tegmark expands on the preceding argument in the following entropic manner. If T differs from 1, the behavior of physical systems could not be predicted reliably from knowledge of the relevant partial differential equations. In such a universe, intelligent life capable of manipulating technology could not emerge. Moreover, if $T > 1$, Tegmark maintains that protons and electrons would be unstable and could decay into particles having greater mass than themselves. (This is not a problem if the particles have a sufficiently low temperature.) If $N > 3$, Ehrenfest's argument above holds; atoms as we know them (and probably more complex structures as well) could not exist. If $N < 3$, gravitation of any kind becomes problematic, and the universe is probably too simple to contain observers. For example, when $N < 3$, nerves cannot cross without intersecting.

In general, it is not clear how physical law could function if T differed from 1. If $T > 1$, subatomic particles which decay after a fixed period would not behave predictably, because time-like geodesics would not be necessarily maximal. $N = 1$ and $T = 3$ has the peculiar property that the speed of light in a vacuum is a *lower bound* on the velocity of matter; all matter consists of tachyons.

Hence entropic and other arguments rule out all cases except $N = 3$ and $T = 1$—which happens to describe the world about us. Curiously, the cases $N = 3$ or 4 have the richest and most difficult geometry and topology. There are, for example, geometric statements whose truth or falsity is known for all N except one or both of 3 and 4. $N = 3$ was the last case of the Poincaré conjecture to be proved.

For an elementary treatment of the privileged status of $N = 3$ and $T = 1$ of Barrow; for deeper treatments, of Barrow and Tipler (1986) and Tegmark. Barrow has repeatedly cited the work of Whitrow.

String theory hypothesizes that matter and energy are composed of tiny vibrating strings of various types, most of which are embedded in dimensions that exist only on a scale no larger than the Planck length. Hence $N = 3$ and $T = 1$ do not characterize string theory, which embeds vibrating strings in coordinate grids having 10, or even 26, dimensions.

Conclusion: The space-time continuum confirms the unity of the center of the energy who supplies all systems (solar systems, galaxies, animals, plants, human beings, Jenny, evils and angels) with the repeating of the unit energy and the space to form the space-time continuum as one unit, since the time or the continuation of all the systems in the space of the universe coming from the repetition of the energy units in all the systems by the center of the energy of the universe.

General point of view

The existence depends mainly on:

1— The center of energy of the universe who communicates with all the systems in the universe to continue their existence in the space which can be called time. The special theory of relativity tells us that matter is a sort of energy by the Law $E = mc^2$, where m is the mass of the matter and c is the light velocity, 3×10^5 km s^{-1}. So, the center of energy of the universe is the source of creation of the matter from the energy and communicates with all the systems inside the universe space to give them the continuation in the space.

2— The space in which all the materials are created by the center of energies and to be existed in the universe is the theater of the life. Space contains the matter such as electrons, positrons, protons, neutrons or more generally nucleons, solar systems, galaxies, animals, plants, light, evils, angels, fields (electric, magnetic fields) and electromagnetic waves. This means that the center of energies and the space is one unit since the space is the theater of creation for the center of the energies; this can be called life-system. This system of the existence does not leave us to choose his name. Since the human being is the relation between the space-time and the matter, and has a relation with the energies center of the universe, therefore the system of the existence puts his name in the right-hand of the human being in

Arabic language such as Allah الله.

The name of Allah in Arabic language
In the right hand of human being.
(Right hand of the reader)

From religious point of view, Holy Quran confirms that Allah has a face that has all kinds of powers and energies.

Holy Quran tells us in Surah Al-rahman (26-27) that every thing will be finally destroyed at the end of the universe except the glorious face of Allah. This can be concluded as:

Allah Possesses All Kinds of Power

This may tell us that the glorious face of Allah possesses all kinds of powers and energies to give the continuation of all the systems from atoms to galaxies and all kinds of animals, human beings, Jenny, evils and angels [Every Thing].

3— This orients our mind to the place from which Allah, God of all, calls his slaves from the human beings and Jenny to believe in him from a country talking Arabic language namely El-Hegaz (old name) or kingdom of Saudi Arabia, especially from the place in which house of Allah exists i.e. Mecca (origin of Arabic language). But who is this prophet whom has been chosen by Allah. Look to the human being that has been created in the skeleton of the name of Mohammed in Arabic language such as follows:

Name of Prophet Mohamed in Arabic Language
(Man has been constructed in this skeleton)

Since the human being has been created in the skeleton of the name of the prophet Mohammed and has the name of system of existence, Allah in Arabic language in his right-hand, therefore man occupies the earth planet as successor of Allah in the earth. Therefore the name of Allah in the right hand of the human being preserves man from predator animals to live and populates in the earth planet.

When Moslems Stand To Pray:

The prayers of Moslems start praying by standing keeping their heads in a definite position in the space, and then they kneel in front of them between their arms. Then they return to their back in the space to retain their stand position again. Finally they prostrate putting the foreheads touching the ground in another position in the space on the land toward Al-Kabba which is filled with the space. In general point of view, Moslem prayers take different positions in the space forward and backward in the orientation toward Al-Kabba, house of Allah, which had been built by Prophet Abraham and his son Ismaiel.

Religious Point of View:

1—   From Holy Quran we know that Allah has right hand catching every thing in the universe as well as earth planet and sun etc . . . This can be explained by the space overall the universe surrounding everything.

Holy Quran in surah Tabarak-1 tells us that Allah has the universe in his hands; this Surah verifies that Allah has every thing in his hand.

Also in Surah yaseen-83 Quran tells that everything is in the hand of Allah and all human beings finally return to Allah.

Also Holy Quran in surah Al-zomor-67 tells us that all the sky is in his right-hand of Allah. This surah tells us that Allah possesses every thing in his hand that can be realized or illustrated by the Space.

The Space has been mentioned in Holy Quran in Surah Al-Waqi'ah 74-76. In this Surah Allah swears in the Holy Quran with the positions of stars and Allah didn't swear with the stars themselves. Of course positions of stars can be determined relatively by the Cartesian coordinates; this means that the positions are the space in which the stars move in their orbitals. In other words, every point in the space is a position i.e. The space is positions. This is the only swearing that has been glorified by Allah in the Holy Quran.

In Surah Al-Aaraf-57 in the Holly Quran, Allah tells us that he sends the wind as indication between his clement hand to derive the heavy clouds to poor land. Of course, the wind contains air molecules moving in the Space that mentioned in the Holy Quran as the clement hands.

Also Allah in Surah Yaseen-45 asked us in the Holy Quran to glorify what is in front of us and what is in our back. What is between our hands always? Simply air molecules existing in the Space in front of us and behind us to give us the freedom to move everywhere and breathing air which is very important for every animal and plants. Motion of human beings is the replacement of the air molecules in the space.

Also Holy Quran in Surah Al-Naml 8-9 told that when Prophet Moses went to the fire on the mountain in cold whether to be warmed during his return to Egypt with his wife, he heard a voice glorifies what is inside and what is outside surrounding the fire. What is inside and what is outside surrounding the fire? Of course fire spread out in a part of the space therefore space is inside and is outside surrounding the fire. Then Allah said that I am the honorable and the wise Allah, God of All. From religious point of view the voice glorifies the right hand of Allah which is inside and outside surrounding the fire. Why was the fire? The fire is the symbol of power and energy in the right hand of Allah.

Holy Quran in Surah Sabaa-9, Allah gives strongly attention for what is between the sky and the earth and this is miracle for a repentant person who wants worshiping his God. What is between the earth and sky? Space is the main thing filling between them.

Holy Quran in Surah Yoones-61, Allah told us that Allah is deponent for our reading Holy Quran (talking) and working where we move in what is deponent on us. As well, He told us that He is not languishes to define any atomic weight in the earth or in the sky. We are talking, working and moving in the Space.

Holy Quran in Surah Al-Nesaa-108, Allah told us that He is ambient with our working. Of course, our motion in the Space is the replacement of air molecules in the Space.

2—   When Moses went to talk to Allah in the mountain he asked him to see his glory (his face). In the holy Bible—Exodus 33: 18-23 (18) Then Moses asked to see God's glory (19) The Lord replied" I will make my goodness pass before you, and I will announce to you the meaning of my name Jehovah, the Lord. I show kindness and mercy to any one I want to. (20) But you may not see the glory of my face, for man may not see me and live. (21) However, stand here on this rock beside me. (22) And when my glory goes by, I will put you in the cleft of the rock and cover you with my hand until I have passed. (23) Then I will remove my hand and you shall see my back, but not my face.

    This is the situation between Allah and Moses when Moses asked to see the face of Allah in the Holy Bible. But from Holy Quran point of view in Surah Al-Aaraf-143, Prophet Moses asked Allah to see him. But Allah told him "you will not see me". Then Allah wanted to teach Moses how he couldn't see Allah who told Moses "If the mountain persists to the appearance of Allah, Moses will see him. But after the appearance of the face of Allah to the mountain, the mountain sank immediately in the land then Moses fell down being unconscious. This can be explained simply that the mountain had been exposed to very high energy from infinite distance associated with the appearance of the face of Allah who possesses all kind of the power of the universe and this verifies that Allah who talked to Moses is the God of the entire universe.

Also Holy Quran in Surah Al-Rahman 26-27 states always that Allah has a face. In this surah in Holy Quran tells us that every thing will be destroyed except the face of Allah.

Also in Surah Al-Bakrah-115 Allah tells us that Allah has the sunrises and the sunsets in addition, the face of Allah exists in everywhere you can look. Also, this Surah tells that Allah is wide and omniscient.

Also Allah points out to his face in surah Al-kasas-88, telling us that every thing going to be destroyed except his face and we return to him: As well he denies calling other God with him.

3— Mohammed is the great prophet of Allah and all the Allah books told about him even in the English Holy Bible in Deuteronomy-18 telling us that "(18) And I gave them other instruction at that time also. (19)(20)(21) Then we left Mount Horeb and traveled through the great and terrible desert, finally arriving among the Amorite hills to which the Lord our God had directed us. We were then at Kadesh—barnea [on the border of Promised Land] and I said to the people, "The Lord God has given us this land. Go and possess it as he told us to. Don't be afraid. Don't even doubt. But in the Arabic Holy Bible told us that Allah constructs to them a prophet among his brothers like Moses and Allah Puts his words in his mouth. **This prophet talks with the orders of Allah and Allah will punish persons who don't obey him.** Also the English Holy Bible tells us in Isaiah 29:12, 13, that" So all of these future events are sealed book to them. When you give it to one who can read, he says I can't, for it is sealed. When you give to another, he says "Sorry I can't read". This remembers us when Prophet Mohammed was in cavern Heraa in Mecca, in KSA, the Holy Spirit, Gabriel, descended to him and ordered Prophet Mohammed to read in the name of generous Allah but Mohammed told him "I can't read, I can't read". Mohamed was illiterate and he heard Holy Quran from The Holy Spirit, Gabriel, to tell all people the words of Holy Quran.

Also Allah tells us in Holy Quran in Surah Al-Aaraf-157: that "who follow the illiterate Prophet that is found in the Bible ". This Surah verifies that Mohammed is illiterate Prophet as he had been in his life.

Some Future Events in Holy Quran

1-invasion of the space

In Surah Al-rahman-33-35: Allah told us about the invasion of the space. He challenged both of Jenny and Man to penetrate the Space except by sultan (Power). Yes Allah has told us in the Holy Quran that Man will invade the space by sultan since 1431 years when man has no any idea about invasion of the space even the earth planet. As well when both Jenny and Man arrive to the horizons of the space they will be thrown by fire and copper and they will not triumph.

Also Holy Quran in Surah Fosselat (52-54) told us since 1431 years that Allah will show whom don't believe in his words in Holy Quran the miracles in the horizons (space) and in themselves to be sure that Allah is the truth and it is enough that Allah is deponent on everything. Yes they suspect in Allah but he is circumambient with every thing.

Race of Space

After the Soviet space program's launch of the world's first artificial satellite (*Sputnik 1*) on October 4, 1957, the attention of the United States turned toward its own fledgling space efforts. The U.S. Congress, alarmed by the perceived threat to national security and technological leadership (known as the "Sputnik crisis"), urged immediate and swift action; President Dwight D. Eisenhower and his advisers counseled more deliberate measures. Several months of debate produced an agreement that a new federal agency was needed to conduct all non-military activity in space. The Advanced

Research Projects Agency (ARPA) was also created at this time to develop space technology for military application.

From late 1957 to early 1958, the National Advisory Committee for Aeronautics (NACA) began studying what a new non-military space agency would entail, as well as what its role might be, and assigned several committees to review the concept. On January 12, 1958, NACA organized a "Special Committee on Space Technology", headed by Guy ford Stever. Stever's committee included consultation from the Army Ballistic Missile Agency's large booster program, referred to as the Working Group on Vehicular Program, headed by Wernher von Braun, a German scientist who became a naturalized US citizen after World War II.

On January 14, 1958, NACA Director Hugh Dryden published "A National Research Program for Space Technology" stating:

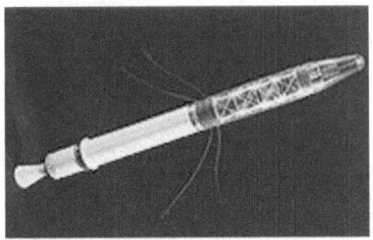

Explorer 1, first US satellite

Launched on January 31, 1958, Explorer 1, officially Satellite 1958 Alpha, became the U.S.'s first earth satellite. The Explorer 1 payload consisted of the Iowa Cosmic Ray Instrument without a tape data recorder which was not modified in time to make it onto the satellite.

On March 5, PSAC Chairman James Killian wrote a memorandum to President Eisenhower, entitled "Organization for Civil Space Programs", encouraging the creation of a civil space program based upon a "strengthened and re-designated" NACA which could expand its research program "with a minimum of delay." In late March, a NACA report entitled "Suggestions for a Space Program" included recommendations for subsequently developing a hydrogen fluorine fueled rocket of 4,450,000 Newton (1,000,000 lb$_f$) thrust designed with second and third stages.

In April 1958, Eisenhower delivered to the U.S. Congress an executive address favoring a national civilian space agency and submitted a bill to create a "National Aeronautical and Space Agency." NACA's former role of research alone would change to include large-scale development, management, and operations. The U.S. Congress passed the bill, somewhat reworded, as the National Aeronautics and Space Act of 1958, on July 16. Only two days later von Braun's Working Group submitted a preliminary report severely criticizing the duplication of efforts and lack of coordination among various organizations assigned to the United States' space programs. Stever's Committee on Space Technology concurred with the criticisms of the von Braun Group (a final draft was published several months later, in October).

On July 29, 1958, Eisenhower signed the National Aeronautics and Space Act, establishing NASA. When it began operations on October 1, 1958, NASA absorbed the 46-year-old NACA intact; its 8,000 employees, an annual budget of US$100 million, three major research laboratories (Langley Aeronautical Laboratory, Ames Aeronautical Laboratory, and Lewis Flight Propulsion Laboratory) and two small test facilities.

Elements of the Army Ballistic Missile Agency, of which von Braun's team was a part, and the Naval Research Laboratory were incorporated into NASA. A significant contributor to NASA's entry into the Space Race with the Soviet Union has the technology from the German rocket program (led by von Braun) which in turn incorporated the technology of Robert Goddard's earlier works. Earlier research efforts within the U.S. Air Force and many of ARPA's early space programs were also transferred to NASA. In December 1958, NASA gained control of the Jet Propulsion Laboratory, a contractor facility operated by the California Institute of Technology.

May 5, 1961 launch of Redstone rocket and *Freedom 7*
with Alan Shepard on first US manned sub-orbital spaceflight

Project Mercury

Conducted under the pressure of the competition between the U.S. and the Soviet Union that existed during the Cold War, Project Mercury was initiated in 1958 and started NASA down the path of human space exploration with missions designed to discover if man could survive in space. Representatives from the U.S. Army, Navy, and Air Force were selected to provide assistance to NASA. Pilot selections were facilitated through coordination with U.S. defense research, contracting, and military test pilot programs. On May 5, 1961, astronaut Alan Shepard became the first American in space when he piloted *Freedom 7* on a 15-minute suborbital flight. John Glenn became the first American to orbit the Earth on February 20, 1962 during the flight of *Friendship 7*. Three more orbital flights followed.

Project Gemini

Launch of Gemini 1

Project Gemini focused on conducting experiments and developing and practicing techniques required for lunar missions. The first Gemini flight with astronauts on board, Gemini 3, was flown by Gus Grissom and John Young on March 23, 1965. Nine missions followed, showing that long-duration human space flight and rendezvous and docking with another vehicle in space were possible, and gathering medical data on the effects of weightlessness on humans. Gemini missions included the first American spacewalks, and new orbital maneuvers including rendezvous and docking.

The Apollo 11-Saturn V space vehicle lifts off on July 16, 1969.

Program of Apollo

The Apollo program landed the first humans on Earth's Moon. Apollo 11 landed on the moon on July 20, 1969 with astronauts Neil Armstrong and Buzz Aldrin, while Michael Collins orbited above. Five subsequent Apollo missions also landed astronauts on the Moon, the last in December 1972. In these six Apollo spaceflights twelve men walked on the Moon. These missions returned a wealth of scientific data and 381.7 kilograms (842 lb) of lunar samples. Experiments included soil mechanics, meteoroids, seismic, heat flow, lunar ranging, magnetic fields, and solar wind experiments.

Apollo 11 Lunar Module Pilot Buzz Aldrin salutes US flag

Apollo set major milestones in human spaceflight. It stands alone in sending manned missions beyond low Earth orbit, and landing humans on another celestial body. Apollo 8 was the first manned spacecraft to orbit another celestial body, while Apollo 17 marked the last moonwalk and the last manned mission beyond low Earth orbit. The program spurred advances in many areas of technology peripheral to rocketry and manned spaceflight, including avionics, telecommunications, and computers. Apollo sparked interest in many fields of engineering and left many physical facilities and machines developed for the program as landmarks. Many objects and artifacts from the program are on display at various locations throughout the world, notably at the Smithsonian's Air and Space Museums.

NASA's Skylab space station

Skylab was the first space station the United States launched into orbit. The 100 short tons (91 t) station was in Earth orbit from 1973 to 1979, and was visited by crews three times, in 1973 and 1974. It included a laboratory for studying the effects of microgravity, and a solar observatory. A Space Shuttle was planned to dock with and elevate Skylab to a higher safe altitude, but Skylab reentered the atmosphere and was destroyed in 1979, before the first shuttle could be launched.

Moon Voyage

Aldrin took this picture of Armstrong in the cabin after the completion of the EVA.

During the *Apollo 11* launch, Armstrong's heart reached a top rate of 109 beats per minute. He found the first stage to be the loudest—much noisier than the *Gemini 8* Titan II launch—and the Apollo CSM was relatively roomy compared to the confinement of the Gemini capsule. This ability to move around was suspected to be the cause of space sickness that had hit members of previous crews, but none of the *Apollo 11* crew suffered from it. Armstrong was especially happy, as he had been prone to motion sickness as a child and could experience nausea after doing long periods of aerobatics.

The objective of *Apollo 11* was to land safely rather than touch down with precision on a particular spot. Three minutes into the lunar descent burn he noted that craters were passing about two seconds too early, which meant the *Eagle* would likely land beyond the planned landing zone by several miles. As the *Eagle's* landing radar acquired the surface, several computer error alarms appeared. The first was a code 1202 alarm and even with their extensive training Armstrong or Aldrin were not aware of what this code meant. However, they promptly received word from CAPCOM in Houston that the alarms were not a concern. The 1202 and 1201 alarms were caused by an executive overflow in the lunar module computer. As described by Buzz Aldrin in the documentary *In the Shadow of the Moon*, the overflow condition was caused by his own counter-checklist choice of leaving the docking radar on during the landing process, so the computer had to process unnecessary radar data and did not have enough time to execute all tasks, dropping lower-priority ones. Aldrin stated that he did so with the objective of facilitating re-docking with the CM should an abort become necessary, not realizing that it would cause the overflow condition.

First Moon Landing issue commemorating *Apollo 11*.
Armstrong is not honored "by portrayal" in accordance with USPS
criteria pertaining to postage issues not honoring living people.

Armstrong took over manual control of the LM, found an area which to him seemed safe for a landing and touched down on the moon at 20:17:39 UTC on July 20, 1969. Some accounts of the *Apollo 11* landing describe the LM's fuel situation as having been dire, with only a few seconds remaining when they touched down. Armstrong had landed the LLTV with less than 15 seconds left on several occasions and he was also confident the LM could survive a straight-down fall from 50 feet (15 m) if needed. Analysis after the mission showed that because of the moon's lower gravity, fuel had sloshed about in the tank more than anticipated, which led to a misleadingly low indication of the remaining propellant; at touchdown there were about 50 seconds of propellant burn time left.

When a sensor attached to the legs of the still hovering Lunar Module made lunar contact, a panel light inside the LM lit up and Aldrin called out, "Contact light." As the LM settled on the surface Aldrin then said, "Okay. Engine stop," and Armstrong said, "Shutdown." The first words Armstrong intentionally spoke to Mission Control and the world from the lunar surface were, "Houston, Tranquility Base here. The *Eagle* has landed". Aldrin and Armstrong celebrated with a brisk handshake and pat on the back before quickly returning to the checklist of tasks needed to ready the lunar module for liftoff from the Moon should an emergency unfold during the first moments on the lunar surface. During the critical landing, the only message from Houston was *"30 seconds"*, meaning the amount of fuel left. When Amstrong had confirmed touch down, Houston expressed their worries during the manual landing as *"You got a bunch of guys about to turn blue. We're breathing again".*

First walk on Moon

Although the official NASA flight plan called for a crew rest period before extra-vehicular activity, Armstrong requested that the EVA be moved earlier in the evening, Houston time. Once Armstrong and Aldrin were ready to go outside, *Eagle* was depressurized, the hatch was opened and Armstrong made his way down the ladder first.

At the bottom of the ladder, Armstrong said "I'm going to step off the LEM now" (referring to the Apollo Lunar Module). He then turned and set his left boot on the surface at 2:56 UTC July 21, 1969. Then spoke the famous words "That's one small step for [a] man, one giant leap for mankind."

Neil Armstrong

Armstrong had decided on this statement following a train of thought that he had had after launch and during the hours after landing. Speaking the line, he accidentally dropped the "a", from his remark, rendering the phrase a contradiction (as *man* in such use is synonymous with *mankind*). Armstrong later said he "would hope that history would grant me leeway for dropping the syllable and understand that it was certainly intended, even if it was not said—although it might actually have been." It has since been claimed that acoustic analysis of the recording reveals the presence of the missing "a". A digital audio analysis conducted by Peter Shann Ford, an Australia-based computer programmer, claims that Armstrong did, in fact, say "a man", but the "a" was inaudible due to the limitations of communications technology of the time. Ford and James R. Hansen, Armstrong's authorized biographer, presented these findings to Armstrong and NASA representatives, who conducted their own analysis. The article by Ford, however, is published on Ford's own web site rather than in a peer-reviewed scientific journal, and linguists David Beaver and Mark Liberman at *Language Log* were skeptical of Ford's claims. Armstrong has expressed his preference that written quotations include the "a" in parentheses.

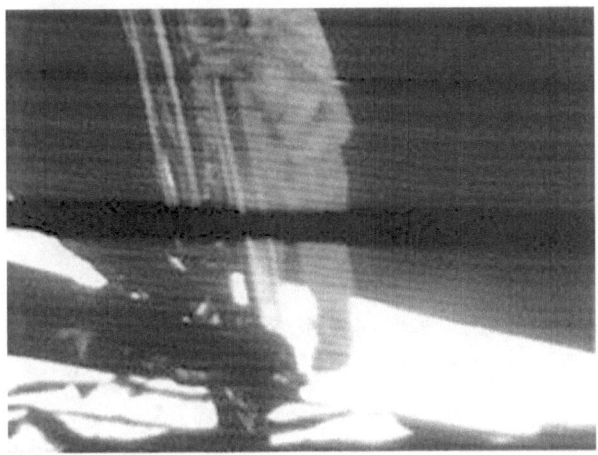

Armstrong prepares to take the first step on the Moon.

When Armstrong made his proclamation, Voice of America was rebroadcast live via the BBC and many other stations the world over. The global audience at that moment was estimated at 450 million listeners, out of a then estimated world population of 3.631 billion people.

About 15 minutes after the first step, Aldrin joined Armstrong on the surface and became the second human to set foot on the Moon. The duo began their tasks of investigating how easily a person could operate on the lunar surface. Early on they also unveiled a plaque commemorating their flight, and also planted the flag of the United States. The flag used on this mission had a metal rod to hold it horizontal from its pole. Since the rod did not fully extend, and the flag was tightly folded and packed during the journey, the flag ended up with a slightly wavy appearance, as if there were a breeze. On Earth there had been some discussion as to whether it was appropriate to plant the flag at all. Armstrong has said that he personally did not think that any flag should have been left, but decided it wasn't worth making a big deal about. Slayton had warned Armstrong that they would receive a special communication, but did not tell him that President Richard Nixon would contact them just after the flag planting.

Armstrong works at the Apollo Lunar Module in one of the
few photos showing him during the EVA.

In the entire *Apollo 11* photographic record, there are only five images of Armstrong partly shown or reflected. The mission was planned to the minute, with the majority of photographic tasks to be performed by Armstrong with their single Hasselblad camera. Aldrin has explained that there were plans to take a photo of Armstrong after the famous image of Aldrin was taken, but they were interrupted by the Nixon communication, which began just five minutes later.

After helping to set up the Early Apollo Scientific Experiment Package, Armstrong went for a walk to what is now known as East Crater, 65 yards (60 m) east of the LM, the greatest distance traveled from the LM on the mission. Armstrong's final task was to leave a small package of memorial items to deceased Soviet cosmonauts Yuri Gagarin and Vladimir Komarov, and Apollo 1 astronauts Gus Grissom, Ed White and Roger Chaffee. The time spent on EVA during *Apollo 11* was about two-and-a-half hours, the shortest of any of the six Apollo lunar landing missions. Each of the subsequent five landings were allotted gradually longer periods for EVA activities. The crew of *Apollo 17*, by comparison, spent over 21 hours exploring the lunar surface.

Return to the Earth

After they re-entered the LM, the hatch was closed and sealed. While preparing for the liftoff from the lunar surface, Armstrong and Aldrin discovered that in their bulky spacesuits, they had broken the ignition switch for the ascent engine. The ascent engine had no switch to fire. Using part of a pen, they pushed the circuit breaker in to activate the launch sequence. Aldrin still possesses the pen which they used to do this. (Aldrin has it kept in a glass case for all to see). The lunar module then continued to its rendezvous and docked with *Columbia*, the command and service module, and returned to Earth. The command module splashed down in the Pacific Ocean and the Apollo 11 crew was picked up by the USS *Hornet* (CV-12).

The *Apollo 11* crew and President Richard Nixon.

After being released from an 18-day quarantine to ensure that they had not picked up any infections or diseases from the Moon, the crew was feted across the United States and around the world as part of a 45-day "Giant Leap" tour. Armstrong then took part in Bob Hope's 1969 USO show, primarily to Vietnam.

In May 1970, Armstrong traveled to the Soviet Union to present a talk at the 13th annual conference of the International Committee on Space Research. Arriving in Leningrad from Poland, he traveled to Moscow where he met Premier Alexei Kosygin. He was the first westerner to see the supersonic Tupolev Tu-144 and was given a tour of the Yuri Gagarin Cosmonauts Training Center, which Armstrong described as "a bit Victorian in nature." At the end of the day, he was surprised to view delayed video of the launch of Soyuz 9. It had not occurred to Armstrong that the mission was taking place, even though Valentina Tereshkova had been his host and her husband, Andriyan Nikolayev, was on board.

ASTP

The Apollo-Soyuz Test Project (ASTP) was the first joint flight of the U.S. and Soviet space programs. The mission took place in July 1975. For the United States, it was the last Apollo flight, as well as the last manned space launch until the flight of the first Space Shuttle in April 1981.

Program of Space Shuttle

The first space shuttle launch, April 12, 1981
Main article: Space Shuttle program

The Space Shuttle became the major focus of NASA in the late 1970s and the 1980s. Planned as a frequently launch-able and mostly reusable vehicle, four space shuttle orbiters were built by 1985. The first to launch, *Columbia*, did so on April 12, 1981.

In 1995 Russian-American interaction resumed with the Shuttle-Mir missions. Once more an American vehicle docked with a Russian craft, this time a full-fledged space station. This cooperation continues to today, with Russia and America the two biggest partners in the largest space station ever built: the International Space Station (ISS). The strength of their cooperation on this project was even more evident when NASA began relying on Russian launch vehicles to service the ISS during the two-year grounding of the shuttle fleet following the 2003 Space Shuttle *Columbia* disaster.

The shuttle fleet lost two orbiters and 14 astronauts in two disasters: *Challenger* in 1986, and *Columbia* in 2003.[27] While the 1986 loss was mitigated by building the Space Shuttle *Endeavour* from replacement parts, NASA did not build another orbiter to replace the second loss. NASA's shuttle program has made 132 successful launches as of May 2010.

The International Space Station

The International Space Station (ISS) is an internationally developed research facility currently being assembled in Low Earth Orbit. On-orbit construction of the station began in 1998 and is scheduled to be completed by 2011, with operations continuing until at least 2015. The station can be seen from the Earth with the naked eye, and, as of 2009, is the largest artificial satellite in Earth orbit, with a mass larger than that of any previous space station.

The ISS is operated as a joint project among NASA, the Russian Federal Space Agency (RKA), the Japan Aerospace Exploration Agency (JAXA), the Canadian Space Agency (CSA), and the European Space Agency (ESA). Ownership and utilization of the space station is set out via several intergovernmental treaties and agreements, with the Russian Federation retaining full ownership of its own modules, and the rest of the station being allocated among the other international partners. The International Space Station relied on the Shuttle fleet for all major construction shipments.

The cost of the station project has been estimated by ESA as €100 billion over a course of 30 years, although cost estimates vary between 35 billion dollars and 160 billion dollars, making the ISS the most expensive object ever constructed.

Mariner program

Picture of Mariner 6

The Mariner program conducted by NASA launched a series of robotic interplanetary probes designed to investigate Mars, Venus and Mercury. The program included a number of firsts, including the first planetary flyby, the first pictures from another planet, the first planetary orbiter, and the first gravity assist maneuver.

Of the ten vehicles in the Mariner series, seven were successful and three were lost. The planned Mariner 11 and Mariner 12 vehicles evolved into Voyager 1 and Voyager 2 of the Voyager program, while the Viking 1 and Viking 2 Mars orbiters were enlarged versions of the Mariner 9 spacecraft. Other Mariner-based spacecraft, launched since Voyager, included the Magellan probe to Venus, and the Galileo probe to Jupiter. A second-generation Mariner spacecraft, called the Mariner Mark II series, eventually evolved into the Cassini-Huygens probe, now in orbit around Saturn.

All Mariner spacecraft were based on a hexagonal or octagonal "bus", which housed all of the electronics, and to which all components were attached, such as antennae, cameras, propulsion, and power sources. All probes except Mariner 1, Mariner 2 and Mariner 5 had TV cameras. The first five Mariners were launched on Atlas-Agena rockets, while the last five used the Atlas-Centaur. All Mariner-based probes after Mariner 10 used the Titan IIIE, Titan IV unmanned rockets or the Space Shuttle with a solid-fueled Inertial Upper Stage and multiple planetary flybys.

Program of Pioneer

Artist's conception of the Pioneer Venus Orbiter

The Pioneer program is a series of NASA unmanned space missions that were designed for planetary exploration. There were a number of such missions in the program, but the most notable were Pioneer 10 and Pioneer 11, which explored the outer planets and left the solar system. Both carry a golden plaque, depicting a man and a woman and information about the origin and the creators of the probes, should any extraterrestrials find them someday.

Additionally, the Pioneer mission to Venus consisted of two components, launched separately. Pioneer Venus 1 or *Pioneer Venus Orbiter* was launched in 1978 and studied the planet for more than a decade after orbital insertion in 1978. Pioneer Venus 2 or *Pioneer Venus Multi-probe* sent four small probes into the Venusian atmosphere.

Program of Voyager

Voyager 1 launch, September 5, 1977

The Voyager program is a series of NASA unmanned space missions that consist of a pair of unmanned scientific probes, *Voyager 1* and *Voyager 2*. They were launched in 1977 to take advantage of a favorable planetary alignment of the late 1970s. Although they were officially designated to study just Jupiter and Saturn, the two probes were able to continue their mission into the outer solar system. Both probes have achieved escape velocity from the solar system and will never return. Both missions have gathered large amounts of data about the gas giants of the solar system, of which little was previously known.

Voyager 1 is currently the farthest human-made object from Earth at about 110.94 AU (16.596 billion kilometers (**Expression error: Missing operand for * mi**), or 10.312 billion miles), traveling away from both the Earth and the Sun at a speed of 17 kilometers (11 mi)/s, which corresponds to a greater specific orbital energy than any other probe.

Program of Viking

Dr. Carl Sagan With the full-scale model of the Viking Lander

The Viking program consisted of a pair of space probes sent to Mars—Viking 1 and Viking 2. Each vehicle was composed of two main parts, an orbiter designed to photograph the surface of Mars from orbit, and a Lander designed to study the planet from the surface. The orbiters also served as communication relays for the Landers once they touched down. Viking 1 was launched on August 20, 1975, and the second craft, Viking 2, was launched on September 9, 1975, both riding atop Titan III-E rockets with Centaur upper stages. By discovering many geological forms that are typically formed from large amounts of water, the Viking program caused a revolution in scientific ideas about water on Mars.

The primary objectives of the Viking orbiters were to transport the Landers to Mars, perform reconnaissance to locate and certify landing sites, act as a communications relays for the Landers, and to perform their own scientific investigations. The orbiter, based on the earlier Mariner 9 spacecraft, was an octagon approximately 2.5 m across. The total launch mass was 2,328 kilograms (5,130 lb), of which 1,445 kilograms (3,190 lb) were propellant and attitude control gas.

Helios probes

Prototype of the Helios spacecraft

The Helios I and Helios II space probes, also known as Helios-A and Helios-B, were a pair of probes launched into heliocentric orbit for the purpose of studying solar processes. A joint venture of the Federal Republic of Germany (West Germany) and NASA, the probes were launched from Cape Canaveral Air Force Station, Florida, on Dec. 10, 1974, and Jan. 15, 1976, respectively. The probes are notable for setting a maximum speed record among spacecraft at 252,792 kilometers (157,078 mi)/h (157,078 mi/h or 43.63 mi/s or 70.22 kilometers (43.63 mi)/s or 0.000234c). The Helios space probes completed their primary missions by the early 1980s, but they continued to send data up to 1985. The probes are no longer functional but still remain in their elliptical orbit around the Sun.

Hubble Space Telescope

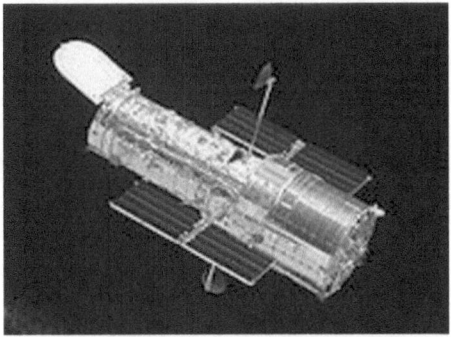

The Hubble Space Telescope

The Hubble Space Telescope (HST) is a space telescope that was carried into orbit by the space shuttle in April 1990. It is named after the American astronomer Edwin Hubble. Although not the first space telescope, Hubble is one of the largest and most versatile, and is well-known as both a vital research tool and a public relations boon for astronomy. The HST is a collaboration between NASA and the European Space Agency, and is one of NASA's Great Observatories, along with the Compton Gamma Ray Observatory, the Chandra X-ray Observatory, and the Spitzer Space Telescope.[30] The HST's success has paved the way for greater collaboration between the agencies.

The HST was created with a relatively small budget of $2 billion and has continued operation since 1990, delighting both scientists and the public. Some of its images, such as the groundbreaking Hubble Deep Field, have become famous.

Magellan probe

The Magellan Probe prepared for launch

The Magellan spacecraft was a space probe sent to the planet Venus, the first unmanned interplanetary spacecraft to be launched by NASA since its successful Pioneer Orbiter, also to Venus, in 1978. It was also the first of three deep-space probes to be launched on the Space Shuttle, and the first spacecraft to employ aero-braking techniques to lower its orbit.

Magellan created the first (and currently the best) high resolution mapping of the planet's surface features. Prior Venus missions had created low resolution radar globes of general, continent-sized formations. Magellan, performed detailed imaging and analysis of craters, hills, ridges, and other geologic formations, to a degree comparable to the visible-light photographic mapping of other planets.

Galileo probe

The Galileo probe

Galileo was an unmanned spacecraft sent by NASA to study the planet Jupiter and its moons. It was launched on October 18, 1989 by the Space Shuttle *Atlantis* on the STS-34 mission. It arrived at Jupiter on December 7, 1995, a little more than six years later, via gravitational assist flybys of Venus and Earth.

Despite antenna problems, *Galileo* conducted the first asteroid flyby, discovered the first asteroid moon, was the first spacecraft to orbit Jupiter, and launched the first probe into Jupiter's atmosphere. Galileo's prime mission was a two-year study of the Jovian system. The spacecraft traveled around Jupiter in elongated ellipses, each orbit lasting about two months. The differing distances from Jupiter afforded by these orbits allowed *Galileo* to sample different parts of the planet's extensive magnetosphere. The orbits were designed for close up flybys of Jupiter's largest moons. Once Galileo's prime mission was concluded, an extended mission followed starting on December 7, 1997; the spacecraft made a number of daring close flybys of Jupiter's moons Europe and Io. The closest approach was 180 kilometers (110 mi) (112 mi) on October 15, 201.

On September 21, 2003, after 14 years in space and eight years of service in the Jovian system, *Galileo*'s mission was terminated by sending the orbiter into Jupiter's atmosphere at a speed of nearly 50 kilometers per second to avoid any chance of it contaminating local moons with bacteria from Earth. Of particular interest was the ice-crusted moon Europe, which, thanks to *Galileo*, scientists now suspect harbors a salt water ocean beneath its surface.

Mars Global Surveyor

Artist's conception of the Mars Global Surveyor

The *Mars Global Surveyor* (MGS) was developed by NASA's Jet Propulsion Laboratory and launched November 1996. It began the United States's return to Mars after a 10-year absence. It completed its primary mission in January 2001 and was in its third extended mission phase when, on November 2, 2006, the spacecraft failed to respond to commands. In January 2007 NASA officially ended the mission.

The *Surveyor* spacecraft used a series of high-resolution cameras to explore the surface of Mars during its mission, returning more than 240,000 images spanning portions of 4.8 Martian years, from September 1997 to

November 2006. The surveyor's cameras utilized 3 instruments: a narrow angle camera that took (black-and-white) high resolution images (usually 1.5 to 12 m per pixel) red and blue wide angle pictures for context (240 m per pixel) and daily global imaging (7.5 kilometers (4.7 mi) per pixel).

The Soujourner rover on Mars

Mars Pathfinder

The *Mars Pathfinder* (MESUR Pathfinder), later renamed the *Carl Sagan Memorial Station*, was launched on December 4, 1996, just a month after the *Mars Global Surveyor* was launched. Onboard the Lander was a small rover called *Sojourner* that would execute many experiments on the Martian surface. It was the second project from NASA's Discovery Program, which promotes the use of low-cost spacecraft and frequent launches under the motto "cheaper, faster and better" promoted by the then administrator, Daniel Goldin. The mission was directed by the Jet Propulsion Laboratory, a division of the California Institute of Technology, responsible for NASA's Mars Exploration Program.

This mission, besides being the first of a series of missions to Mars that included rovers (robotic exploration vehicles), was the most important since the *Vikings* landed on the red planet in 1976, and also was the first successful mission to send a rover to a planet. In addition to scientific objectives, the Mars Pathfinder mission was also a "proof-of-concept" for various technologies, such as airbag-mediated touchdown and automated obstacle avoidance, both later exploited by the Mars Exploration Rovers. The Mars Pathfinder was also remarkable for its extremely low price relative to other unmanned space missions to Mars.

Artist's conception of MER on Mars

Mars Exploration Rovers

NASA's *Mars Exploration Rover Mission* (MER), is an ongoing robotic space mission involving two rovers exploring the planet Mars. The mission is managed for NASA by the Jet Propulsion Laboratory, which designed, built and is operating the rovers.

The mission began in 2003 with the sending of the two rovers—MER-A *Spirit* and MER-B *Opportunity*—to explore the Martian surface and geology. The mission's scientific objective is to search for and characterize a wide range of rocks and soils that hold clues to past water activity on Mars. The mission is part of NASA's Mars Exploration Program which includes three previous successful Landers: the two Viking program Landers in 1976 and Mars Pathfinder probe in 1997.

The total cost of building, launching, landing and operating the rovers on the surface for the initial 90-Martian-day (sol) primary mission was US$820 million. Since the rovers have continued to function far beyond their initial 90 sol primary mission (to date both rovers have been functioning on Mars's surface for nearly seven years), they have each received multiple mission extensions. In recognition of the vast amount of scientific information amassed by both rovers, two asteroids have been named in their honor: 37452 Spirit and 39382 Opportunity

New Horizons probe

Artist's conception of New Horizons orbiting Pluto

New Horizons is a NASA robotic spacecraft mission currently en route to the dwarf planet Pluto. It is expected to be the first spacecraft to fly by and study Pluto and its moons, Charon, Nix, and Hydra. Once New Horizons leaves the Solar System, NASA may also approve flybys of one or more other Kuiper Belt Objects.

New Horizons was launched on January 19, 2006 directly into an Earth-and-solar-escape trajectory. It had an Earth-relative velocity of about 16.26 kilometers (10.10 mi)/s or 58,536 kilometers (36,373 mi)/h (10.10 mi/s or 36,373 mi/h) after its last engine shut down. Thus, it left Earth at the fastest launch speed ever recorded for a man-made object (although it's specific orbital energy is less than that of Voyager 1, and the Helios Probes retain the maximum speed record for a spacecraft). New Horizons flew by Jupiter on February 28, 2007 and Saturn's orbit on June 8, 2008. It will arrive at Pluto on July 14, 2015 and then continue into the Kuiper belt.

NASA's future

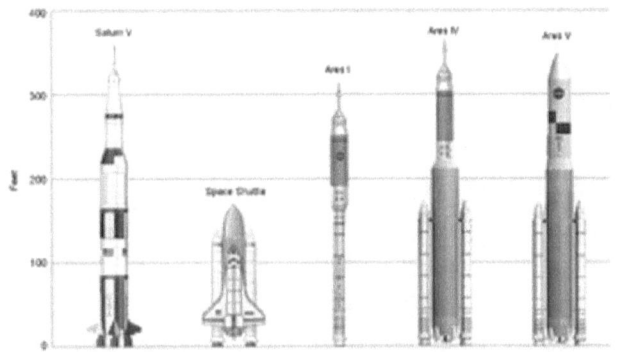

Left to Right: Saturn V, which carried men to the Moon,
the Space Shuttle, and the proposed Ares I, Ares IV and Ares V
launch vehicles

During much of the 1990s, NASA was faced with shrinking annual budgets due to congressional belt-tightening. In response, NASA's ninth administrator, Daniel Goldin, pioneered the "faster, better, cheaper" approach that enabled NASA to cut costs while still delivering a wide variety of aerospace programs (Discovery Program). That method was criticized and re-evaluated following the twin losses of Mars Climate Orbiter and Mars Polar Lander in 1999.

It is the current space policy of the United States that NASA, "execute a sustained and affordable human and robotic program of space exploration and develop, acquire, and use civil space systems to advance fundamental scientific knowledge of our Earth system, solar system, and universe."[37]

NASA's ongoing investigations include in-depth surveys of Mars and Saturn and studies of the Earth and the Sun. Other NASA spacecraft are presently en route to Mercury, Pluto and the asteroid belt. With missions to Jupiter in planning stages, NASA's itinerary covers over half the solar system.

An improved and larger planetary rover, Mars Science Laboratory, is under construction and slated to launch in 2011, after a slight delay caused by hardware challenges, which has bumped it back from the October 2009 scheduled launch. The New Horizons mission to Pluto was launched in 2006 and will fly by Pluto in 2015. The probe received a gravity assist from Jupiter in February 2007, examining some of Jupiter's inner moons and testing on-board instruments during the fly-by. On the horizon of NASA's plans is the MAVEN spacecraft as part of the Mars Scout Program to study the atmosphere of Mars.

Orion contractor selected August 31, 2006, at NASA Headquarters

Vision for Space Exploration

On January 14, 2004, ten days after the landing of the Mars Exploration Rover *Spirit*, US President George W. Bush announced a new plan for NASA's future, dubbed the Vision for Space Exploration. According to this plan, mankind would return to the Moon by 2018, and set up outposts as a tested and potential resource for future missions. The Space Shuttle will be retired in 2010 and Orion may replace it by 2015, capable of both docking with the International Space Station (ISS) and leaving the Earth's orbit. The future of the ISS is somewhat uncertain—construction will be completed, but beyond that is less clear. Although the plan initially met with skepticism from Congress, in late 2004 Congress agreed to provide start-up funds for the first year's worth of the new space vision.

Hoping to spur innovation from the private sector, NASA established a series of Centennial Challenges, technology prizes for non-government teams, in 2004. The Challenges include tasks that will be useful for implementing the Vision for Space Exploration, such as building more efficient astronaut gloves. In February 2010, NASA announced that it would be awarding $50 million in contracts to commercial spaceflight companies including Blue Origin, Boeing, Paragon Space Development Corporation, Sierra Nevada Corporation and United Launch Alliance to design and develop viable reusable launch vehicles

Moon Base

On December 4, 2006, NASA announced it was planning a permanent moon base. NASA Associate Administrator Scott Horowitz said the goal was to start building the moon base by 2020, and by 2024, has a fully functional base that would allow for crew rotations and in-situ resource utilization. Additionally, NASA plans to collaborate and partner with other nations for this project. As of February 1, 2010, however, President Obama has scrapped the possibility of a moon base through his budget as he believes that NASA should be more focused on deep space missions.

Human exploration of Mars

On September 28, 2007 Michael D. Griffin, who was at the time Administrator of NASA, stated that NASA aims to put a man on Mars by 2037.

Alan Stern, NASA's "hard-charging" and "reform-minded" associate administrator for the Science Mission Directorate, resigned on March 25, 2008, effective April 11, 2008, after he allegedly ordered funding cuts to the Mars Exploration Rover (MER) and Mars Odyssey that were overturned by NASA Administrator Michael D. Griffin. The cuts were intended to offset cost overruns for the Mars Science Laboratory. Stern has stated that he "did not quit over MER" and that he "wasn't the person who tried to cut MER". Stern, who served for nearly a year and has been credited with making "significant changes that have helped restore the importance of science in NASA's mission", says he left to avoid cutting healthy programs and basic research in favor of politically sensitive projects. Griffin favored cutting "less popular parts" of the budget, including basic research, and Stern's refusal to do so lead to his resignation.

Recent developments

President Obama and Senator Bill Nelson arrive at
Kennedy Space Center in April 2010.

Boeing and Lockheed Martin have expressed doubts about President Obama's plans to drop the Moon and Mars missions and instead focus on "space taxis" limited to trips to orbital stations such as the ISS, while other aerospace companies including Space X have strongly endorsed the cancellation of the Constellation program and alternative proposals which were announced officially in an April 15, 2010 space policy speech at Kennedy Space Center.

Controversy

The chair and ranking member of the U.S. Senate Committee on Homeland Security and Governmental Affairs wrote NASA Administrator Griffin on July 31, 2006 expressing concerns about the change. NASA also canceled or delayed a number of earth science missions in 2006.

In 2009, NASA announced that the agency plans to provide $1.75 million in funding to Jack Bergman of Harvard's McLean Hospital to conduct an experiment on monkeys to determine the health effects of radiation exposure during travel in deep space. The plan has faced opposition from animal rights groups such as PETA and HSUS, a physicians' group PCRM, and several federal legislators lead by Representative Jim Moran of Virginia who claim that the grant should be cancelled because during the course of the experiment, the primates will likely contract malignant tumors as well as blindness, skin damage, cognitive decline, premature aging and death. PCRM also claims that the proposed use and isolation of primates would violate NASA's stated principles regarding animal ethics. The group has filed a federal complaint alleging that the experiments would also violate the Animal Welfare Act.

Public perception of the NASA budget is very different from reality and has been the subject of controversy since the agency's creation. A 1997 poll reported that Americans had an average estimate of 20% for NASA's share of the federal budget. In reality, NASA's budget has been between 0.5% and 1% from the late 1960's on. NASA budget briefly peaked at over 4% of the federal budget in the mid 1960's during the build up to the Apollo program.

Facilities

NASA headquarters in Washington, DC provides overall guidance and direction to the agency.[68] NASA's Shared Services center is located on the grounds of the John C. Stennis Space Center, near Bay St. Louis, Mississippi. [69] Construction of the Shared Services facility began in August 2006 and it was completed in June 2008. NASA operates a short-line railroad at the Kennedy Space Center. Various field and research installations are listed below by application. Some facilities serve more than one application for historic or administrative reasons.

Research centers

JPL complex in Pasadena, California

- Ames Research Center, Moffett Federal Airfield, Mountain View, California
- Jet Propulsion Laboratory, California Institute of Technology, Pasadena, California
- Goddard Institute for Space Studies, New York City
- Goddard Space Flight Center, Greenbelt, Maryland
- John H. Glenn Research Center at Lewis Field, Cleveland, Ohio
- Langley Research Center, Hampton, Virginia

Construction and launch facilities

The Vehicle Assembly Building and Launch Control
Center at Kennedy Space Center.

- George C. Marshall Space Flight Center, Huntsville, Alabama
- John F. Kennedy Space Center, Florida
- Lyndon B. Johnson Space Center, Houston, Texas
- Michoud Assembly Facility, New Orleans, Louisiana
- Wallops Flight Facility, Wallops Island, Virginia
- White Sands Test Facility, Las Cruces, New Mexico

Deep Space Network

- Deep Space Network (DSN) stations

    ○ Canberra Deep Space Communication Complex, Canberra, Australian Capital Territory
    ○ Goldstone Deep Space Communications Complex, Barstow, California

Madrid Deep Space Communication Complex, Madrid, Spain

Florida, USA, taken from shuttle mission STS-95 on October 31, 1998

Ozone depletion

In the middle of the 20ᵗʰ century,. NASA augmented its mission of Earth's observation and redirected it toward environmental quality. The result was the launch of Earth Observing System (EOS) in 1980s, which was able to monitor one of the global environmental problems—ozone depletion. The first comprehensive worldwide measurements were obtained in 1978 with the Nimbus-7 satellite and NASA scientists at the Goddard Institute for Space Studies.

Salt evaporation and energy management

In one of the nation's largest restoration projects, NASA technology helps state and federal government reclaim 15,100 acres (61 km²) of salt evaporation ponds in South San Francisco Bay. Satellite sensors are used by scientists to study the effect of salt evaporation on local ecology.

NASA has started Energy Efficiency and Water Conservation Program as an agency-wide program directed to prevent pollution and reduce energy and water utilization. It helps to ensure that NASA meets its federal stewardship responsibilities for the environment.

Earth Science Enterprise

Understanding of natural and human-induced changes on the global environment is the main objective of NASA's Earth Science Enterprise. For years it has been cooperating with major environment related agencies and creating united projects to achieve their goal. Past Enterprise's programs include:

- Carbon sequestration assessment for Carbon Management (USDA, DOE)
- Early warning systems for air and water quality for Homeland Security (OHS, NIMA, USGS)
- Enhanced weather prediction for Energy Forecasting (DOE, United States Environmental Protection Agency (EPA))
- Environmental indicators for Coastal Management (NOAA)
- Environmental indicators for Community Growth Management (EPA, USGS, NSGIC)
- Environmental models for Biological Invasive Species (USGS, USDA)
- Regional to national to international atmospheric measurements and predictions for Air Quality Management (United States Environmental Protection Agency, NOAA)
- Water cycle science for Water Management and Conservation (EPA, USDA)

NASA is working in cooperation with National Renewable Energy Laboratory (NREL). The goal is to obtain to produce worldwide solar resource maps with great local detail.[78] NASA was also one of the main participants in the evaluation innovative technologies for the clean up of the sources for dense non-aqueous phase liquids (DNAPLs). On April 6, 1999, the agency signed The Memorandum of Agreement (MOA) along with the United States Environmental Protection Agency, DOE, and USAF authorizing all the above organizations to conduct necessary tests at the John F. Kennedy Space center. The main purpose was to evaluate two innovative in-situ remediation technologies, thermal removal and oxidation destruction of DNAPLs. National Space Agency made a partnership with Military Services and Defense Contract Management Agency named the "Joint Group on Pollution Prevention". The group is working on reduction or elimination of hazardous materials or processes.

On May 8, 2003, Environmental Protection Agency recognized NASA as the first federal agency to directly use landfill gas to produce energy at one of its facilities—the Goddard Space Flight Center, Greenbelt, Maryland.

Other prediction of the Holy Quran is in Surah Al-Araaf-167

2—HOLOCAUST

The Holy Quran has told us since 1431 years in Surah Al-Aaraf 167 that Allah will send who agonizes the Israel people the badest torment till the end of the world and he has quick nemeses as well as he is clement and he has the forgiveness.

Jewish history

From Wikipedia, the free encyclopedia, 17 August 2010

Jewish history is the history of the Jewish people, religion, and culture. Since Jewish history is over four thousand years long and includes hundreds of different populations, any treatment can only be provided in broad strokes.

Ancient Israelites

Moses with the Tablets of Stone
(1659 painting by Rembrandt)

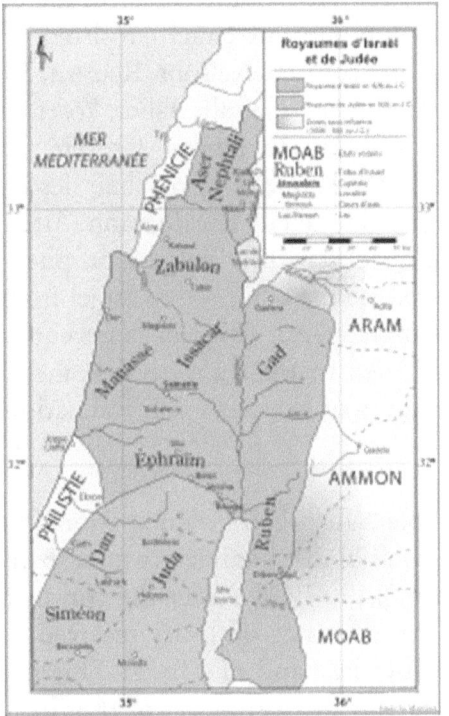

Kingdoms of Israel and Judah in 926 BCE

For the first two periods the history of the Jews is mainly that of the Fertile Crescent. It begins among those people who occupied the area lying between the Nile, Tigris and the Euphrates rivers. Surrounded by ancient seats of culture in Egypt and Babylonia, by the deserts of Arabia, and by the highlands of Asia Minor, the land of Canaan (roughly corresponding to modern-day Israel, the Palestinian Territories, Jordan and Lebanon) was a meeting place of civilizations. The land was traversed by old-established trade routes and possessed important harbors on the Gulf of Akaba and on the Mediterranean coast, the latter exposing it to the influence of other cultures of the Fertile Crescent.

According to the Bible, Jews around the world are descended from the ancient Hebrew people of Israel who settled in the land of Canaan, located between the eastern coast of the Mediterranean Sea and the Jordan River (1451 BCE). The Children of Israel shared a lineage through their common ancestors, Abraham, his son Isaac, and Isaac's son Jacob, Hebrews whose nomadic travels centered around Hebron somewhere between 1991 and 1706 BCE. The Children of Israel consisted of twelve tribes, each descendant from one of Jacob's twelve sons, Reuven, Shimon, Levi, Yehuda, Yissachar, Zevulun, Dan, Gad, Naftali, Asher, Yosef, and Benyamin. Jacob and his twelve sons left Canaan during a severe famine and settled in Goshen of northern Egypt. While in Egypt their descendants were enslaved by the Egyptian government led by the Pharaoh. After 400 years of slavery, YHWH the God of Israel sent the Hebrew prophet Moses, a man from the tribe of Levi, to release the Children of Israel from Egyptian bondage. Israel miraculously emigrated out of Egypt (an event known as the Exodus, in the Bible and returned to their ancestral homeland in Canaan. This event marks the formation of Israel as a political nation in Canaan, in 1400 BCE.

According to the Bible, after their emancipation from Egyptian slavery, the people of Israel dwelt in the Sinai desert for a span of forty years before conquering Canaan in 1400 BCE under the command of Joshua. While living in the desert, the nation of Israel received the Torah at Mount Sinai from YHWH, by the hand of Moses. This marked the beginning of normative Judaism and the formation of the first Abrahamic religion. After entering Canaan, portions of the land were given to each of the twelve tribes of Israel. For several hundred years, Israel was organized into a confederacy of twelve tribes ruled by a series of Judges. In 1000 BCE, an Israelite monarchy was established under Saul, and continued under King David and his son, Solomon. During the reign of David, Jerusalem eternally became the national and spiritual capital of Israel. David's son Solomon built the First Temple on Mount Moriah in Jerusalem. Upon his death a civil war erupted between the ten northern Israelite tribes, and the tribes of Judah and Benjamin in the south. The nation split into two states, Israel, consisting of ten of the tribes (in the north), and the Kingdom of Judah, consisting of the tribes of Judah and Benjamin (in the south). Israel was conquered by the Assyrian ruler Shalmaneser V in the 8th century BCE. There is no commonly accepted historical record of those ten tribes, which are sometimes referred to as the Ten Lost Tribes of Israel.

Babylonian captivity

Deportation and exile of the Jews of the ancient Kingdom of Judah
to Babylon and the destruction of Jerusalem and Solomon's temple

The kingdom of Judah was conquered by a Babylonian army in the early
6th century BCE. The Judahite elite was exiled to Babylon, but later at
least a part of them returned to their homeland, led by prophets Ezra and
Nehemiah, after the subsequent conquest of Babylonia by the Persians.
Since Zoroastrianism was the state religion of the Persian Empire, the
extent to which Zoroastrianism has been an influence in the development
of Judaism is a subject of some debate among scholars.

Post-exilic period

Model of the Second Temple of Jerusalem

Construction of the Second Temple was completed under the leadership of the last three Jewish Prophets Haggai, Zechariah and Malachi with Persian approval and financing. After the death of the last Jewish Prophets and still under Persian rule, the leadership of the Jewish people was in the hands of five successive generations of zugot ("pairs of") leaders. They flourished first under the Persians then under the Greeks. As a result the Pharisees and Sadduccees were formed. Under the Persians then under the Greeks, Jewish coins were minted in Judea as Yehud coinage.

Hellenistic period

In 332 BCE the Persians were defeated by Alexander the Great. After his demise, and the division of Alexander's empire among his generals, the Seleucid Kingdom was formed.

During this time currents of Judaism were influenced by Hellenistic philosophy developed from the 3rd century BCE, notably the Jewish diaspora in Alexandria, culminating in the compilation of the Septuagint. An important advocate of the symbiosis of Jewish theology and Hellenistic thought is Philo.

The Hasmonean Kingdom

A coin (Hendin 485) issued by Mattathias Antigonus circa
40 BCE featuring a Menorah

A deterioration of relations between hellenized Jews and religious Jews
led the Seleucid king Antiochus IV Epiphanes to impose decrees banning
certain Jewish religious rites and traditions. Consequently, the orthodox Jews
revolted under the leadership of the Hasmonean family, (also known as the
Maccabees). This revolt eventually led to the formation of an independent
Jewish kingdom, known as the Hasmonaean Dynasty, which lasted from
165 BCE to 63 BCE. The Hasmonean Dynasty eventually disintegrated as
a result of civil war between the sons of Salome Alexandra, Hyrcanus II and
Aristobulus II. The people, who did not want to be governed by a king but
by theocratic clergy, made appeals in this spirit to the Roman authorities.
A Roman campaign of conquest and annexation, led by Pompey, soon
followed.

Roman Rule in the Land of Israel (63 BCE-324 CE)

Siege and Destruction of Jerusalem by the Romans
(1850 painting by David Roberts)

The sack of Jerusalem depicted on the inside wall of
the Arch of Titus in Rome

Judea under Roman rule was at first an independent Jewish kingdom first by the Hasmonaeans then by the Herodians, but gradually their power declined, until it came under the direct rule of Romans and renamed the *Iudaea Province*. The empire was often callous and brutal in its treatment of its Jewish subjects. In 66 CE, the Jews began to revolt against the Roman rulers of Judea. The revolt was defeated by the future Roman emperors Vespasian and Titus. In the Siege of Jerusalem in 70 CE, the Romans destroyed much of the Temple in Jerusalem and, according to some accounts, plundered artifacts from the temple, such as the Menorah. Jews continued to live in their land in significant numbers, the Kitos War of 115-117 CE nothwithstanding, until Julius Severus ravaged Judea while putting down the Bar Kokhba revolt of 132-136 CE. 985 villages were destroyed and most of the Jewish population of central Judaea was essentially wiped out, killed, sold into slavery, or forced to flee[citation needed]. Banished from Jerusalem, the Jewish population now centred on Galilee.

The diaspora

Many of the Judaean Jews were sold into slavery while others became citizens of other parts of the Roman Empire. The book of Acts in the New Testament, as well as other Pauline texts, make frequent reference to the large populations of Hellenised Jews in the cities of the Roman world. These Hellenised Jews were only affected by the diaspora in its spiritual sense, absorbing the feeling of loss and homelessness which became a cornerstone of the Jewish creed, much supported by persecutions in various parts of the world. The policy towards proselytism and conversion to Judaism, which spread the Jewish religion throughout the Hellenistic civilization, seems to have ended with the wars against the Romans and the following reconstruction of Jewish values for the post-Temple era.

Of critical importance to the reshaping of Jewish tradition from the Temple-based religion to the traditions of the Diaspora, was the development of the interpretations of the Torah found in the Mishnah and Talmud.

Late Roman Period in the Land of Israel

In spite of the failure of the Bar Kokhba revolt, Jews remained in the land of Israel in significant numbers. The Jews who remained there went through numerous experiences and armed conflicts against consecutive occupiers of the Land. Some of the most famous and important Jewish texts were composed in Israeli cities at this time. The Jerusalem Talmud, the completion of the Mishnah and the system of niqqud are examples.

In this period the *tannaim* and *amoraim* were active, rabbis who organized and debated the Jewish oral law. The decisions of the *tannaim* are contained in the Mishnah, Beraita, Tosefta, and various Midrash compilations. The Mishnah was completed shortly after 200 CE, probably by Judah haNasi. The commentaries of the *amoraim* upon the Mishnah are compiled in the Jerusalem Talmud, which was completed around 400 CE, probably in Tiberias.

In 351 CE, the Jewish population in Sepphoris, under the leadership of Patricius, started a revolt against the rule of Constantius Gallus, brother-in-law of Emperor Constantius II. The revolt was eventually subdued by Gallus' general, Ursicinus.

According to tradition, in 359 CE Hillel II created the Hebrew calendar based on the lunar year. Until then, the entire Jewish community outside the land of Israel depended on the calendar sanctioned by the Sanhedrin; this was necessary for the proper observance of the Jewish holy days. However, danger threatened the participants in that sanction and the messengers who communicated their decisions to distant communities.

As the religious persecutions continued, Hillel determined to provide an authorized calendar for all time to come.

In 363, shortly before launching his campaign against the Sassanid Empire, the last pagan Roman Emperor, Julian II, allowed the Jews to return to "holy Jerusalem which you have for many years longed to see rebuilt" and to rebuild the Temple. However, Julian's campaign against the Persians failed and he was killed in battle on 26 June 363. The Temple was not rebuilt.

Middle Ages
Byzantine period in the land of Israel
(324-638)

Jews were widespread throughout the Roman Empire, and this carried on to a lesser extent in the period of Byzantine rule in the central and eastern Mediterranean. The militant and exclusive Christianity and caesaropapism of the Byzantine Empire did not treat Jews well, and the condition and influence of diaspora Jews in the Empire declined dramatically.

It was official Christian policy to convert Jews to Christianity, and the Christian leadership used the official power of Rome in their attempts. In 351 CE the Jews revolted against the added pressures of their Governor, one named Gallus. Gallus put down the revolt and destroyed the major cities in the Galilee where the revolt had started. Tzippori and Lydda (site of two of the major legal academies) never recovered.

Nonetheless it is in this period that the Nasi in Tiberias, Hillel II created an official calendar which needed no monthly sightings of the moon. The months were set, and the calendar needed no further authority from Judea. At about the same time, the Jewish academy at Tiberius began to collate the combined Mishnah, braitot, explanations, and interpretations developed by generations of scholars who studied after the death of Judah HaNasi. The text was organized according to the order of the Mishna: each paragraph of Mishnah was followed by a compilation of all of the interpretations, stories, and responses associated with that Mishnah. This text is called the Jerusalem Talmud.

The Jews of Judea received a brief respite from official persecution during the rule of the Emperor Julian the Apostate. Julian's policy was to return the kingdom to Hellenism and he encouraged the Jews to rebuild Jerusalem. Julian's rule lasted only from 361 to 363, so there was no chance to carry out this promise before Christian rule was restored over the Empire. Beginning in 398 with the consecration of St. John Chrysostom as Patriarch, the Christian rhetoric against Jews continued to rise with a series of sermons such as "Against the Jews" and "On the Statues, Homily 17" where John preaches against "the Jewish sickness". Such heated language would build a climate of distrust and hate of the large Jewish settlements, such as those in Antioch and Constantinople.

In the beginning of the fifth century, the Emperor Theodosius issued a set of decrees which established official prosecution against Jews. Jews were not allowed to own slaves, build new synagogues, hold public office or try cases between a Jew and a non-Jew. Intermarriage between Jew and non-Jew was made a capital offense as was a Christian converting to Judaism. Theodosius, furthermore, did away with the Sanhedrin and abolished the post of Nasi. Under the Emperor Justinian the authorities restricted the civil rights of Jews, and threatened their religious privileges. The emperor also interfered in the internal affairs of the synagogue, and forbade, for instance, the use of the Hebrew language in divine worship. The recalcitrant were menaced with corporal penalties, exile, and loss of property. The Jews at Borium, not far from Syrtis Major, who resisted the Byzantine General Belisarius in his campaign against the Vandals, were forced to embrace Christianity and their synagogue was converted to a church.

Justinian and his successors of course had concerns outside the province of Judea, and there were insufficient troops to enforce these regulations. As a result, ironically, the sixth century saw a wave of new synagogues built with beautiful mosaic floors. Jews assimilated into their lives the rich art forms of the Byzantine culture. There exist mosaics showing people, animals, menorahs, zodiacs, and Biblical characters. Excellent examples of these synagogue floors have been found at Beit Alpha (which includes the scene of Abraham sacrificing a ram instead of his son Isaac along with a gorgeous zodiac), Tiberius, Beit Shean, and Tzippori.

The precarious existence of Jews under Byzantine rule did not long endure, largely for the explosion of the Muslim religion out of the remote Arabian peninsula (where large populations of Jews resided.

The Muslim Caliphate ejected the Byzantines from the Holy Land (or the Levant, defined as modern Israel, Jordan, Lebanon and Syria) within a few years of their victory at the Battle of Yarmouk in 636. A testament of the cruelty of the Byzantines towards the Jews can be noted in the great number of Jews who fled remaining Byzantine territories in favour of residence in the Caliphate over the subsequent centuries.

Yet, the size of the Jewish community in the Byzantine Empire was not affected by attempts by some emperors (most notably Justinian) to forcibly convert the Jews of Anatolia to Christianity, as these attempts met with very little success. The exact picture of the status of the Jews in Asian Minor during the Byzantine rule is still being researched by historians (for a sample of views, see, for instance, J. Starr "The Jews in the Byzantine Empire, 641-1204", S. Bowman, "The Jews of Byzantium", R. Jenkins "Byzantium", Averil Cameron, "Byzantines and Jews: Recent Work on Early Byzantium," Byzantine and Modern Greek Studies 20). Although there is some evidence of occasional hostility by the Byzantine populations and authorities, no systematic persecution of the type endemic at that time in Western Europe (pogroms, the stake, mass expulsions etc.) has been recorded in Byzantium. Much of the Jewish population of Constantinople remained in place after the conquest of the city by Mehmet II.

A curious historical event did occur as a result of this emigration. Sometime in the 7th or 8th century, the Khazars, a Turkic tribe in what is now the Ukraine, seems to have converted to Judaism. The completeness of this conversion is unclear, but certainly there had been a Jewish population in the Crimea since the Hellenistic era, and these may have been reinforced by Jews leaving the fickle Byzantine governance. Influenced and threatened as they were by both Islam and the Byzantine Empire, and receiving much tangible benefit from their Jewish population, it is speculated that Khazar rulers converted to Judaism in an effort to remain neutral as a safeguard to their independence.

Islamic period in the land of Israel (638-1099)

In 638 CE the Byzantine Empire lost control of the Levant. The Arab Islamic Empire under Khalipha Omar conquered Jerusalem and the lands of Mesopotamia, Syria, Palestine and Egypt. Under the various regimes the Jews suffered massacres and were forced to flee the inland villages towards the coast. They were subsequently induced to return inland after the coastal towns had been destroyed. Nevertheless, the Jews still controlled much of the commerce in Palestine. According to Arab geographer Al-Muqaddasi, the Jews worked as "the assayers of coins, the dyers, the tanners and the bankers in the community." During the Fatimid period, many Jewish officials served in the regime. Professor Moshe Gil documents that at the time of the Arab conquest in 7th century CE, the majority of the population was Jewish.

Crusaders Period in the Land of Israel (1099-1260)

Capture of Jerusalem, 1099

Jews (identifiable by the Judenhut they were required to wear) were massacred by Christian knights during the First Crusade in France and Germany as illustrated in this French Bible from 1250.

In 1099, along with the other inhabitants of the land, the Jews vigorously defended Jerusalem against the Crusaders. When the city fell, the Crusaders gathered them in a synagogue and set it alight. In Haifa, the Jews almost single-handedly defended the town against the Crusaders, holding out for a whole month, (June-July 1099). At this time there were Jewish communities scattered all over the country, including Jerusalem, Tiberias, Ramleh, Ashkelon, Caesarea, and Gaza. Jews were not allowed to hold land in the Crusader period but concentrated their efforts on the commerce in the coastal towns during times of quiescence. Most of them were artisans: glassblowers in Sidon, furriers and dyers in Jerusalem.

During this period, the Masoretes of Tiberias established the Hebrew language orthography, or *niqqud*, a system of diacritical vowel points used in the Hebrew alphabet. A large volume of piyutim and midrashim originated in Palestine at this time.

Maimonides wrote that in 1165 he visited Jerusalem and went up on to the Temple Mount and prayed in the "great, holy house". Maimonides established a yearly holiday for himself and his sons, the 6th of Cheshvan, commemorating the day he went up to pray on the Temple Mount, and another, the 9th of Cheshvan, commemorating the day he merited to pray at the Cave of the Patriarchs in Hebron.

In 1141 Yehuda Halevi issued a call to the Jews to emigrate to the land of Israel and took on the long journey himself. After a stormy passage from Córdoba, he arrived in Egyptian Alexandria, where he was enthusiastically greeted by friends and admirers. At Damietta, he had to struggle against the promptings of his own heart, and the pleadings of his friend Halfon ha-Levi, that he remain in Egypt; and free from intolerant oppression. He started on the tedious land route, trodden of old by the Israelite wanderers in the desert. Again he is met with, worn-out, with broken heart and whitened hair, in Tyre and Damascus. Jewish legend relates that as he came near Jerusalem, over-powered by the sight of the Holy City, he sang his most beautiful elegy, the celebrated "Zionide," "Zion ha-lo Tish'ali." At that instant, he was ridden down and killed by an Arab, who dashed forth from a gate.

Mamluk period in the land of Israel (1260-1517)

In the years 1260-1516, the land of Israel was part of the Empire of the Mamluks who ruled first from Turkey, then from Egypt. War and uprisings, bloodshed and destruction followed Maimonides. Jews suffered persecution and humiliation but the surviving records cite at least 30 Jewish urban and rural communities at the opening of the 16th century.

A notable event during the period was the settlement of Nachmanides in the Old City of Jerusalem in 1267 which since then a continuous Jewish presence existed in Jerusalem until modern day occupation of Jordan in 1948[1]. Nahmanides then settled at Acre, where he was very active in spreading Jewish learning, which was at that time very much neglected in the Holy Land. He gathered a circle of pupils around him, and people came in crowds, even from the district of the Euphrates, to hear him. Karaites were said to have attended his lectures, among them being Aaron ben Joseph the Elder, who later became one of the greatest Karaite authorities. Shortly after his arrival in Jerusalem he addressed a letter to his son Nahman, in which he described the desolation of the Holy City, where there were at that time only two Jewish inhabitants—two brothers, dyers by trade. In a later letter from Acre he counsels his son to cultivate humility, which he considers to be the first of virtues. In another, addressed to his second son, who occupied an official position at the Castilian court, Nahmanides recommends the recitation of the daily prayers and warns above all against immorality. Nahmanides died after having passed the age of seventy-six, and his remains were interred at Haifa, by the grave of Yechiel of Paris. Yechiel emigrated to Acre in 1260, along with his son and a large group of followers. There he established the Tamudic academy *Midrash haGadol d'Paris*. He is believed to have died there between 1265 and 1268.

In 1488 Obadiah ben Abraham, commentator on the Mishnah, arrived in Jerusalem and marked a new epoch for the Jewish community in The Land.

Spain, North Africa, and the Middle East

During the Middle Ages, Jews were generally better treated by Islamic rulers than Christian ones. Despite second-class citizenship, Jews played prominent roles in Muslim courts, and experienced a "Golden Age" in Moorish Spain about 900-1100, though the situation deteriorated after that time. Riots resulting in the deaths of Jews did however occur in North Africa through the centuries and especially in Morocco, Libya and Algeria where eventually Jews were forced to live in ghettos.[14]

The 11th century saw Muslim pogroms against Jews in Spain; those occurred in Cordoba in 1011 and in Granada in 1066. Decrees ordering the destruction of synagogues were enacted in the Middle Ages in Egypt, Syria, Iraq and Yemen. Jews were also forced to convert to Islam or face death in some parts of Yemen, Morocco and Baghdad at certain times. The Almohads, who had taken control of much of Islamic Iberia by 1172, far surpassed the Almoravides in fundamentalist outlook, and they treated the *dhimmis* harshly. Jews and Christians were expelled from Morocco and Islamic Spain. Faced with the choice of either death or conversion, many Jews emigrated. Some, such as the family of Maimonides, fled south and east to the more tolerant Muslim lands, while others went northward to settle in the growing Christian kingdoms.[18][19]

Europe

Jewish populations had existed in Europe, especially in the area of the former Roman Empire, from very early times, with converts to Judaism joined by traders and later by member of the exodus. There are records of Jewish communities in France (see History of the Jews in France) and Germany (see History of the Jews in Germany) from the 4th century, and substantial Jewish communities in Spain even earlier.

Norman Cantor and other twentieth century historians dispute the conventional idea that the Middle Ages was a uniformly difficult time for Jews. Early medieval society, before the Church became fully organized, was tolerant. Between 800 and 1100 there were 1.5 million Jews in Christian Europe. They were fortunate in not being part of the feudal system as serfs or knights, thus were spared the oppression and constant warfare that made life miserable for most Christians. In relations with the Christian society, they were protected by kings, princes and bishops, because of the crucial services they provided in three areas: financial, administrative and as doctors[1]. Christian scholars interested in the Bible would even consult with Talmudic rabbis. All this changed with the reforms and strengthening of the Roman Catholic Church, especially the creations of the Franciscan and Dominican preaching monks, and the rise of envious and competitive middle-class, town-dwelling Christians. By 1300 the friars and local priests were using the Passion Plays at Easter time, which depicted Jews in contemporary dress killing Christ, to teach the general populace to hate and murder Jews. It was at this point that persecution and exile became endemic. Finally around 1500, Jews found security and a renewal of prosperity in Poland.

By and large, Jews were heavily persecuted in Christian Europe after 1300. Since they were the only people allowed to lend money for interest (forbidden to Catholics by the church), some Jews became prominent moneylenders. Christian rulers gradually saw the advantage of having a class of men like the Jews who could supply capital for their use without being liable to excommunication, and the money trade of western Europe by this means fell into the hands of the Jews. However, in almost every instance where large amounts were acquired by Jews through banking transactions the property thus acquired fell either during their life or upon their death into the hands of the king. Jews thus became imperial "servi cameræ," the property of the King, who might present them and their possessions to princes or cities.

According to James Carroll, "Jews accounted for 10% of the total population of the Roman Empire. By that ratio, if other factors had not intervened, there would be 200 million Jews in the world today, instead of something like 13 million."

Jews were frequently massacred and exiled from various European countries. The persecution hit its first peak during the Crusades. In the First Crusade (1096) flourishing communities on the Rhine and the Danube were utterly destroyed; see German Crusade, 1096. In the Second Crusade (1147) the Jews in France were subject to frequent massacres. The Jews were also subjected to attacks by the Shepherds Crusades of 1251 and 1320. The Crusades were followed by expulsions, including in, 1290, the banishing of all English Jews; in 1396, 100,000 Jews were expelled from France; and, in 1421 thousands were expelled from Austria. Many of the expelled Jews fled to Poland.

Early Modern Period
Ottoman period in the land of Israel

Jews lived in the geographic area of Asia Minor (modern Turkey, but more geographically either Anatolia or Asia Minor) for more than 2,400 years. Initial prosperity in Hellenistic times faded under Christian Byzantine rule, but recovered somewhat under the rule of the various Muslim governments which displaced and succeeded rule from Constantinople. For much of the Ottoman period, Turkey was a safe haven for Jews fleeing persecution, and it continues to have a small Jewish population today.

At the time of the Battle of Yarmuk when the Levant passed under Muslim Rule, thirty Jewish communities existed in Haifa, Sh'chem, Hebron, Ramleh, Gaza, Jerusalem, and many in the north. Safed became a spiritual centre for the Jews and the Shulchan Aruch was compiled there as well as many Kabbalistic texts. The first Hebrew printing press, and the first printing in Western Asia began in 1577.

The situation where Jews both enjoyed cultural and economical prosperity at times but were widely persecuted at other times was summarized by G.E. Von Grunebaum :

"It would not be difficult to put together the names of a very sizeable number of Jewish subjects or citizens of the Islamic area who have attained to high rank, to power, to great financial influence, to significant and recognized intellectual attainment; and the same could be done for Christians. But it would again not be difficult to compile a lengthy list of persecutions, arbitrary confiscations, attempted forced conversions, or pogroms."

Historian Martin Gilbert writes that in the 19th century the position of Jews worsened in Muslim countries.[1]

There was a massacre of Jews in Baghdad in 1828.[23] In 1839, in the eastern Persian city of Meshed, a mob burst into the Jewish Quarter, burned the synagogue, and destroyed the Torah scrolls. It was only by forcible conversion that a massacre was averted. There was another massacre in Barfurush in 1867.

In 1840, the Jews of Damascus were falsely accused of having murdered a Christian monk and his Muslim servant and of having used their blood to bake Passover bread or Matza. A Jewish barber was tortured until he "confessed"; two other Jews who were arrested died under torture, while a third converted to Islam to save his life. Throughout the 1860s, the Jews of Libya were subjected to what Gilbert calls punitive taxation. In 1864, around 500 Jews were killed in Marrakech and Fez in Morocco. In 1869, 18 Jews were killed in Tunis, and an Arab mob looted Jewish homes and stores, and burned synagogues, on Jerba Island. In 1875, 20 Jews were killed by a mob in Demnat, Morocco; elsewhere in Morocco, Jews were attacked and killed in the streets in broad daylight. In 1891, the leading Muslims in Jerusalem asked the Ottoman authorities in Constantinople to prohibit the entry of Jews arriving from Russia. In 1897, synagogues were ransacked and Jews were murdered in Tripolitania.[24]

Benny Morris writes that one symbol of Jewish degradation was the phenomenon of stone-throwing at Jews by Muslim children. Morris quotes a 19th century traveler: "I have seen a little fellow of six years old, with a troop of fat toddlers of only three and four, teaching [them] to throw stones at a Jew, and one little urchin would, with the greatest coolness, waddle up to the man and literally spit upon his Jewish gaberdine. To all this the Jew is obliged to submit; it would be more than his life was worth to offer to strike a Mahommedan."[23]

According to Mark Cohen in *The Oxford Handbook of Jewish Studies*, most scholars conclude that Arab anti-Semitism in the modern world arose in the nineteenth century, against the backdrop of conflicting Jewish and Arab nationalism, and was imported into the Arab world primarily by nationalistically minded Christian Arabs (and only subsequently was it "Islamized").[25]

Europe

During the European Renaissance, the worst of the expulsions occurred following the reconquista of Andalus, as the Moorish or Arab Islamic government of Spain was known. With the ejection of the last Muslim rulers from Grenada in 1492, the Spanish Inquisition followed and the entire Spanish population of around 200,000 Sephardic Jews were expelled. This was followed by expulsions in 1493 in Sicily (37,000 Jews) and Portugal in 1496. The expelled Spanish Jews fled mainly to the Ottoman Empire, Holland, and North Africa, others migrating to Southern Europe and the Middle East.

In the 17th century, almost no Jews lived in Western Europe. The relatively tolerant Poland had the largest Jewish population in Europe, but the calm situation for the Jews there ended when Polish and Lithuanian Jews were slaughtered in the hundreds of thousands by the cossacks during Chmielnicki uprising (1648) and by the Swedish wars (1655). Driven by these and other persecutions, Jews moved back to Western Europe in the 17th century. The last ban on Jews (by the English) was revoked in 1654, but periodic expulsions from individual cities still occurred, and Jews were often restricted from land ownership, or forced to live in ghettos.

With the Partition of Poland in the late 18th century, the Jewish population was split between the Russian Empire, Austro-Hungary, and Prussia, which divided Poland for themselves.

The European Enlightenment and Haskalah (18th century)

During the period of the European Renaissance and Enlightenment, significant changes were happening within the Jewish community. The Haskalah movement paralleled the wider Enlightenment, as Jews began in the 18th century to campaign for emancipation from restrictive laws and integration into the wider European society. Secular and scientific education was added to the traditional religious instruction received by students, and interest in a national Jewish identity, including a revival in the study of Jewish history and Hebrew, started to grow. Haskalah gave birth to the Reform and Conservative movements and planted the seeds of Zionism while at the same time encouraging cultural assimilation into the countries in which Jews resided. At around the same time another movement was born, one preaching almost the opposite of Haskalah, Hasidic Judaism. Hasidic Judaism began in the 18th century by Rabbi Israel Baal Shem Tov, and quickly gained a following with its more exuberant, mystical approach to religion. These two movements, and the traditional orthodox approach to Judaism from which they spring, formed the basis for the modern divisions within Jewish observance.

At the same time, the outside world was changing, and debates began over the potential emancipation of the Jews (granting them equal rights). The first country to do so was France, during the French Revolution in 1789. Even so, Jews were expected to integrate, not continue their traditions. This ambivalence is demonstrated in the famous speech of Clermont-Tonnerre before the National Assembly in 1789:

"We must refuse everything to the Jews as a nation and accord everything to Jews as individuals. We must withdraw recognition from their judges; they should only have our judges. We must refuse legal protection to the maintenance of the so-called laws of their Judaic organization; they should not be allowed to form in the state either a political body or an order. They must be citizens individually. But, some will say to me, they do not want to be citizens. Well then! If they do not want to be citizens, they should say so, and then, we should banish them. It is repugnant to have in the state an association of non-citizens, and a nation within the nation . . ."

19th century

NAPOLÉON LE GRAND,
rétablit le culte des Israélites, le 30 Mai 1806.

An 1806 French print depicts Napoleon Bonaparte
emancipating the Jews.

Though persecution still existed, emancipation spread throughout Europe in the 19th century. Napoleon invited Jews to leave the Jewish ghettos in Europe and seek refuge in the newly created tolerant political regimes that offered equality under Napoleonic Law (see Napoleon and the Jews). By 1871, with Germany's emancipation of Jews, every European country except Russia had emancipated its Jews.

Despite increasing integration of the Jews with secular society, a new form of anti-Semitism emerged, based on the ideas of race and nationhood rather than the religious hatred of the Middle Ages. This form of anti-Semitism held that Jews were a separate and inferior race from the Aryan people of Western Europe, and led to the emergence of political parties in France, Germany, and Austria-Hungary that campaigned on a platform of rolling back emancipation. This form of anti-Semitism emerged frequently in European culture, most famously in the Dreyfus Trial in France. These persecutions, along with state-sponsored pogroms in Russia in the late 19th century, led a number of Jews to believe that they would only be safe in their own nation. See Theodor Herzl and History of Zionism.

During this period, Jewish migration to the United States (see American Jews) created a large new community mostly freed of the restrictions of Europe. Over 2 million Jews arrived in the United States between 1890 and 1924, most from Russia and Eastern Europe.

20th century
Modern Zionism

Theodor Herzl, visionary of the Jewish State, in 1901.

During the 1870s and 1880s the Jewish population in Europe began to more actively discuss immigration back to Israel and the re-establishment of the Jewish Nation in its national homeland, fulfilling the biblical prophecies relating to Shivat Tzion. In 1882 the first Zionist settlement—Rishon LeZion—was founded by immigrants who belonged to the "Hovevei Zion" movement. Later on, the "Bilu" movement established many other settlements in the land of Israel.

The Zionist movement was founded officially after the Kattowitz convention (1884) and the World Zionist Congress (1897), and it was Theodor Herzl who began the struggle to get the world superpowers to establish a state for the Jews.

After the First World War, it seemed that the conditions to establish such a state had arrived: The United Kingdom captured Palestine from the Ottoman Empire, and the Jews received the promise of a "National Home" from the British in the form of the Balfour Declaration of 1917, given to Chaim Weizmann.

In 1920 the British Mandate of Palestine began and the British had promised to create and foster a Jewish national home in Palestine. In the beginning, The pro-Jewish Herbert Samuel was appointed High Commissioner in Palestine, the Hebrew University of Jerusalem was established and several big Jewish immigration waves to Palestine occurred; the situation seemed to be going well. Nevertheless. The Arab inhabitants of Palestine were not fond of the increasing Jewish immigration, and began to oppose Jewish settlement and the pro-Jewish policy of the British government by means of violent uprising and terror.

Arab gangs began performing terror acts and murders on convoys and on the Jewish population. After the 1920 Arab riots and 1921 Jaffa riots, the Jewish leadership in Palestine believed that the British had no desire to confront local Arab gangs over their attacks on Palestinian Jews. Realizing that they could not rely on the British administration for protection from these gangs, the Jewish leadership created the Haganah organization to protect their farms and Kibbutzim.

Major riots occurred during the Arab massacres of 1929 and the 1936-1939 Arab revolt in Palestine.

Due to the Arab violence the United Kingdom gradually started to backtrack from the original idea of a Jewish state and to speculate on a binational solution or an Arab state that would have a Jewish minority.

Meanwhile, the Jews of the United States and Europe gained great success in the fields of the science, culture and the economy. The most prominent physicists of Europe during that period were Jews, most notably Albert Einstein. In the Soviet Union, many Jews were involved in the October Revolution and belonged to the communist party.

The Holocaust

A boy raises his hands when the Jews leave the bunkers
after the submission of the Warsaw Ghetto Uprising

During World War II the Holocaust occurred, in which
Nazi Germany carried out systematic state-sponsored
extermination (genocide) of approximately six million European Jews.

In 1933, with the rise to power of Adolf Hitler and the Nazi party in Germany, The Jewish situation became more severe. Economic crises, racial anti-Semitic laws, and a fear of an upcoming war led many Jews to flee from Europe to Palestine, to the United States and to the Soviet Union.

In 1939 World War II began and until 1941 Hitler occupied almost all of Europe, including Poland—where millions of Jews were living at that time—and France. In 1941, when the invasion of the Soviet Union began, Hitler ordered the initiation of the Final Solution—an extensive organized operation on an unprecedented scale, aimed at the annihilation of the Jews of Europe and French North Africa. This genocide, in which six million Jews were murdered methodically and with horrifying cruelty, is known as The Holocaust or *Shoah* (Hebrew term). In Poland, more than one million Jews were murdered in gas chambers at the Auschwitz concentration camp alone.

The massive scale of the Holocaust, and the horrors which happened during it, heavily affected the Jewish nation and world public opinion, which only understood the dimensions of the Holocaust after the war. After the war it became clear that it was impossible to leave the Jews in the hands of the nations of the world anymore, and efforts were increased to establish a shelter for the wounded Jewish nation. The Holocaust will be mentioned laterly.

The Establishment of the State of Israel Declaration of Independence of Israel

David Ben-Gurion proclaiming Israeli independence from
the United Kingdom on May 14, 1948

In 1945 the Jewish resistance organizations in Palestine unified and established the Jewish Resistance Movement. The movement began pressing the British authority and avenging the Arab armies whom attacked Jews. There are different opinions on the success of the violent struggle of the divisions, and the disobedience movement eventually stopped in 1946 in the aftermath King David Hotel bombing. The Jewish leadership decided to center the struggle in the illegal immigration to Palestine and began organizing massive amount of Jewish war refugees from Europe, without the approval of the British authorities. This immigration contributed a great deal to the Jewish settlements in Israel in the world public opinion and the British authorities decided to let the United Nations decide upon the fate of Palestine.

On November 29, 1947 the United Nations decided on dividing the Palestinian country into two states: A Jewish state and an Arab state. The Jewish leadership accepted this decision but the Arabs opposed it and started attacking the Jewish settlements, and so the 1948 Arab-Israeli War started.

In the middle of the war, after the last soldiers of the British mandate left Palestine, David Ben-Gurion proclaimed in 1948 the establishment of the Jewish state of Israel. In 1949 the war ended and the state of Israel started building the state and absorbing massive waves of hundreds of thousands of Jewish refugees from all over the world.

Since 1948, Israel has been involved in a series of major military conflicts, including the 1956 Suez War, 1967 Six-Day War, 1973 Yom Kippur War (Tenth of Ramadan), 1982 Lebanon War, and 2006 Israel-Lebanon conflict, as well as a nearly constant series of ongoing minor conflicts to preserve its national interests.

Since 1977, an ongoing and largely unsuccessful series of diplomatic efforts have been initiated by Israel, its neighbors, and other parties, including the United States and the European Union, to bring about a peace process to resolve conflicts between Israel and its neighbors mainly with Anwar Sadat President of Egypt, to discuss mostly over the fate of the Palestinian people and the fate of Arab borders for Jourdany and Syrria in accord of Camb David in USA. Really President Anwar Sadat was the hero of settlment of the peace with Israel represented by Prime Minister Mnahen Beigin at least with Egypt.

21st century

The Western Wall in Jerusalem, 2008

Today (2010), Israel is a parliamentary democracy with a population of over 7.5 million people, of whom about 5.6 million are Jewish.

The largest Jewish communities are in Israel and the United States, with major communities in France, Argentina, Russia, England, and Canada.

For statistics related to modern Jewish demographics see the article *Jewish population*.

The Jewish Autonomous Oblast, created during the Soviet period, continues to be an autonomous oblast of the Russian state. The Chief Rabbi of Birobidzhan, Mordechai Scheiner, says there are 4,000 Jews in the capital city. Governor Nikolay Mikhaylovich Volkov has stated that he intends to, "support every valuable initiative maintained by our local Jewish organizations." The Birobidzhan Synagogue opened in 2004 on the 70th anniversary of the region's founding in 1934

The Holocaust and "Shoah"

From Wikipedia, the free encyclopedia, 15 Augest 2010

"Selection" on the *Judenrampe*, Auschwitz, May/June 1944. To be sent to the right meant slave labor; to the left, the gas chambers. This image shows the arrival of Hungarian Jews from Carpatho-Ruthenia, many of them from the Berehov ghetto. It was taken by Ernst Hofmann or Bernhard Walter of the SS. Courtesy of Yad Vashem.

The Holocaust (from the Greek **Ὁλόκαυστος** [holókaustos]: *hólos*, "whole" and *kaustós*, "burnt"), also known as The Shoah (Hebrew: **השואה**, Romanized *HaShoah*, "calamity"; Yiddish: **חורבן**, Romanized *Churben* or *Hurban*[3], from the Hebrew for "destruction") was the genocide of approximately six million European Jews during World War II, a programme of systematic state-sponsored extermination by Nazi Germany. The genocide of these six million people was a genocide of two-thirds of the population of nine million Jews who had resided in Europe before the Holocaust.

Some scholars maintain that the definition of the Holocaust should also include the Nazis' systematic murder of millions of people in other groups, including ethnic Poles, Romani, Soviet civilians, Soviet prisoners of war, people with disabilities, homosexuals, Jehovah's Witnesses, and other political and religious opponents. By this definition, the total number of Holocaust victims would be between 11 million and 17 million people.

The persecution and genocide were carried out in stages. Legislation to remove the Jews from civil society was enacted years before the outbreak of World War II. Concentration camps were established in which inmates were used as slave labor until they died of exhaustion or disease. Where the Third Reich conquered new territory in eastern Europe, specialized units called Einsatzgruppen murdered Jews and political opponents in mass shootings. Jews and Romani were confined in overcrowded ghettos before being transported by freight train to extermination camps where, if they survived the journey, the majority of them were systematically killed in gas chambers. Every arm of Nazi Germany's bureaucracy was involved in the logistics of the mass murder, turning the country into what one Holocaust scholar has called "a genocidal state". Opinions differ on how much the civilian German population knew of the government conspiracy against the Jewish population. Most historians claim that the civilian population was not aware of the atrocities that happened in the camps. Robert Gellately however claims that the German government openly announced the conspiracy through the media, and that the German people were aware of every aspect of the conspiracy, except for the use of gas chambers.

Etymology and Use of the Term

The term *holocaust* comes from the Greek word *holókauston*, an animal sacrifice offered to a god in which the whole (*holos*) animal is completely burnt (*kaustos*). Its Latin form (*holocaustum*) was first used with specific reference to a massacre of Jews by the chroniclers Roger of Howden and Richard of Devizes in the 1190s. For hundreds of years, the word *holocaust* was used in English to denote massive sacrifices and great slaughters or massacres. During World War II, the word was used to describe Nazi atrocities regardless of whether the victims were Jews or non-Jews. Since the 1960s, the term has come to be used by scholars and popular writers to refer exclusively to the genocide of Jews.

The term entered common parlance after 1978, the year that the popular *Holocaust (TV miniseries)* was broadcast on the American NBC television network. The series proved that the subject matter could have popular appeal, as well as providing a convenient and enduring term.

The biblical word **Shoah** (השואה) (also spelled **Sho'ah** and **Shoa**), meaning "calamity", became the standard Hebrew term for the Holocaust as early as the 1940s. *Shoah* is preferred by many Jews for a number of reasons, including the theologically offensive nature of the word *holocaust*, which they take to refer to the Greek pagan custom.

Historical usage of Holocaust, Shoah, and Final Solution

The word *holocaust* has been used since the 18th century to refer to the violent deaths of a large number of people. For example, Winston Churchill and other contemporaneous writers used it before World War II to describe the Armenian Genocide of World War I. Since the 1950s its use has increasingly been restricted, with its usage now mainly used as a proper noun to describe the Holocaust perpetrated by Nazi Germany.

Holocaust was adopted as a translation of *Shoah*—a Hebrew word connoting catastrophe, calamity, disaster, and destruction—which was used in 1940 in Jerusalem in a booklet called *Sho'at Yehudei Polin*, and translated as *The Holocaust of the Jews of Poland*. *Shoah* had earlier been used in the context of the Nazis as a translation of *catastrophe*. For example, in 1934, when Chaim Weizmann told the Zionist Action Committee that Hitler's rise to power was an "unvorhergesehene Katastrophe, etwa ein neuer Weltkrieg" ("an unforeseen catastrophe, comparable to another world war"), the Hebrew press translated *Katastrophe* as *Shoah*. In the spring of 1942, the Jerusalem historian BenZion Dinur (Dinaburg) used *Shoah* in a book published by the United Aid Committee for the Jews in Poland to describe the extermination of Europe's Jews, calling it a "catastrophe" that symbolized the unique situation of the Jewish people. The word *Shoah* was chosen in Israel to describe the Holocaust, the term institutionalized by the Knesset on April 12, 1951, when it established *Yom Ha-Shoah Ve Mered Ha-Getaot*, the national day of remembrance. In the 1950s, Yad Vashem, the Israel "Holocaust Martyrs' and Heroes' Remembrance Authority" was routinely translating this into English as "the Disaster". At that time, *holocaust* was often used to mean the conflagration of much of humanity in a nuclear war. Since then, Yad Vashem has changed its practice; the word *Holocaust*, usually now capitalized, has come to refer principally to the genocide of the European Jews.

The usual German term for the extermination of the Jews during the Nazi period was the euphemistic phrase *Endlösung der Judenfrage* (the "Final Solution of the Jewish Question"). In both English and German, "Final Solution" is widely used as an alternative to "Holocaust". For a time after World War II, German historians also used the term *Völkermord* ("genocide"), or in full, *der Völkermord an den Juden* ("the genocide of the Jewish people"), while the prevalent term in Germany today is either *Holocaust* or increasingly *Shoah*.

Use of the term Holocaust for Jewish and non-Jewish victims

While the terms *"Shoah"* and *"Final Solution"* always refer to the fate of the Jews during the Nazi rule, the term *"Holocaust"* is sometimes used in a wider sense to describe other genocides of the Nazi and other regimes.

The Columbia Encyclopedia defines *"Holocaust"* as "name given to the period of persecution and extermination of European Jews by Nazi Germany".[22] The Compact Oxford English Dictionary and Microsoft Encarta give similar definitions. The Encyclopædia Britannica defines *"Holocaust"* as "the systematic state-sponsored killing of six million Jewish men, women, and children and millions of others by Nazi Germany and its collaborators during World War II",[3] although the article goes on to say, "The Nazis also singled out the Roma (Gypsies). They were the only other group that the Nazis systematically killed in gas chambers alongside the Jews."

Scholars are divided on whether the term Holocaust should be applied to all victims of the Nazi mass murder campaign, with some using it synonymously with *"Shoah"* or *"Final Solution of the Jewish Question"*, and others including the killing of Romani peoples (Roma and Sinti), Poles, the deaths of Soviet prisoners of war, Slavs, homosexual men, Jehovah's Witnesses, the disabled, and political opponents.

Yehuda Bauer contends that the Holocaust should include only Jews because it was the intent of the Nazis to exterminate all Jews, while the other groups were not to be totally annihilated. Inclusion of non-Jewish victims of the Nazis in the Holocaust is objected to by many persons including Elie Wiesel, and by organizations such as Yad Vashem established to commemorate the victims of the Holocaust. They say that the word was originally meant to describe the extermination of the Jews, and that the Jewish Holocaust was a crime on such a scale, and of such totality and specificity, as the culmination of the long history of European antisemitism, that it should not be subsumed into a general category with the other crimes of the Nazis.

Michael Burleigh and Wolfgang Wippermann maintain that although all Jews were victims, the Holocaust transcended the confines of the Jewish community—other people shared the tragic fate of victimhood. László Teleki applies the term *"Holocaust"* to both the murder of Jews and Romani peoples by the Nazis.[29] In *The Columbia Guide to the Holocaust*, Donald Niewyk and Francis Nicosia use the term to include Jews, Gypsies and the disabled. Dennis Reinhartz has claimed that Gypsies were the main victims of genocide in Croatia and Serbia during the Second World War, and has called this "the Balkan Holocaust 1941-1945".

Sometimes, the term *"Holocaust"* is used to describe events that have no connection with Europe or World War II. According to David Stannard, the *"American Holocaust"* involved killing of an estimated 50-100 million aboriginal people, and continues on a smaller scale throughout the Americas. The *"Rwandan Holocaust"* refers to the Rwanda genocide of 1994. The *"Cambodian Holocaust"* comprises the mass killings by the Khmer Rouge regime in Cambodia. *"African Holocaust"* describes the slave trade and the colonization of Africa, also known as the *Maafa*. Then there is the prospect of *"Nuclear Armageddon"*, also known as *"Nuclear Holocaust"*.

Distinctive Features
Compliance of Germany's Institutions

Ghettos were established in Europe in which Jews were confined before being shipped to extermination camps.

Michael Berenbaum writes that Germany became a "genocidal state."[8] Every arm of the country's sophisticated bureaucracy was involved in the killing process. Parish churches and the Interior Ministry supplied birth records showing who was Jewish; the Post Office delivered the deportation and denaturalization orders; the Finance Ministry confiscated Jewish property; German firms fired Jewish workers and disenfranchised Jewish stockholders; the universities refused to admit Jews, denied degrees to those already studying, and fired Jewish academics; government transport offices arranged the trains for deportation to the camps; German pharmaceutical companies tested drugs on camp prisoners; companies bid for the contracts to build the crematoria; detailed lists of victims were drawn up using the Dehomag (IBM Germany) company's punch card machines, producing meticulous records of the killings. As prisoners entered the death camps, they were made to surrender all personal property, which was carefully catalogued and tagged before being sent to Germany to be reused or recycled. Berenbaum writes that the Final Solution of the Jewish question was "in the eyes of the perpetrators . . . Germany's greatest achievement."

Saul Friedländer writes that: "Not one social group, not one religious community, not one scholarly institution or professional association in Germany and throughout Europe declared its solidarity with the Jews."[35] He writes that some Christian churches declared that *converted* Jews should be regarded as part of the flock, but even then only up to a point.

Friedländer argues that this makes the Holocaust distinctive because antisemitic policies were able to unfold without the interference of countervailing forces of the kind normally found in advanced societies, such as industry, small businesses, churches, and other vested interests and lobby groups.

Dominance of ideology and the scale of the genocide

In other genocides, pragmatic considerations such as control of territory and resources were central to the genocide policy. Yehuda Bauer argues that:

The basic motivation [of the Holocaust] was purely ideological, rooted in an illusionary world of Nazi imagination, where an international Jewish conspiracy to control the world was opposed to a parallel Aryan quest. No genocide to date had been based so completely on myths, on hallucinations, on abstract, nonpragmatic ideology—which was then executed by very rational, pragmatic means."

Responding to the German philosopher Ernst Nolte who claimed that the Holocaust was not unique, the German historian Eberhard Jäckel wrote in 1986 that the Holocaust was unique because:

"the National Socialist killing of the Jews was unique in that never before had a state with the authority of its responsible leader decided and announced that a specific human group, including its aged, its women and its children and infants, would be killed as quickly as possible, and then carried through this resolution using every possible means of state power".

The slaughter was systematically conducted in virtually all areas of Nazi-occupied territory in what are now 35 separate European countries. [38] It was at its worst in Central and Eastern Europe, which had more than seven million Jews in 1939. About five million Jews were killed there, including three million in occupied Poland and over one million in the

Soviet Union. Hundreds of thousands also died in the Netherlands, France, Belgium, Yugoslavia and Greece. The Wannsee Protocol makes clear that the Nazis also intended to carry out their "final solution of the Jewish question" in England and Ireland.

Anyone with three or four Jewish grandparents was to be exterminated without exception. In other genocides, people were able to escape death by converting to another religion or in some other way assimilating. This option was not available to the Jews of occupied Europe, unless their grandparents had converted prior to January 18, 1871. All persons of recent Jewish ancestry were to be exterminated in lands controlled by Germany.

Medical experiments

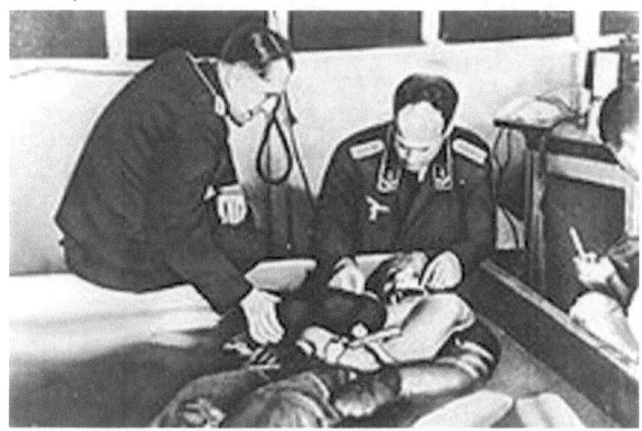

A cold water immersion experiment at Dachau concentration camp presided over by Professor Holzlohner (left) and Dr. Rascher (right)

Another distinctive feature of the Holocaust was the extensive use of human subjects in medical experiments. German physicians carried out such experiments at Auschwitz, Dachau, Buchenwald, Ravensbrück, Sachsenhausen and Natzweiler concentration camps.

The most notorious of these physicians was Dr. Josef Mengele, who worked in Auschwitz. His experiments included placing subjects in pressure chambers, testing drugs on them, freezing them, attempting to change eye color by injecting chemicals into children's eyes and various amputations and other brutal surgeries. The full extent of his work will never be known because the truckload of records he sent to Dr. Otmar von Verschuer at the Kaiser Wilhelm Institute was destroyed by von Verschuer. Subjects who survived Mengele's experiments were almost always killed and dissected shortly afterwards.

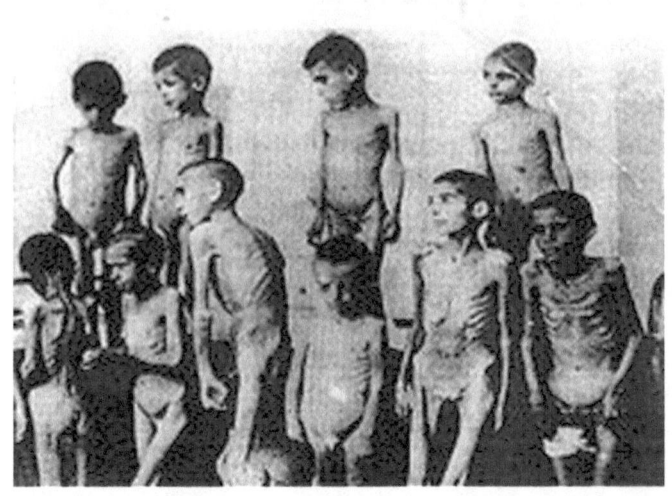

Romani children in Auschwitz, victims of medical experiments

He seemed particularly keen on working with Romani children. He would bring them sweets and toys, and personally take them to the gas chamber. They would call him "Onkel Mengele". Vera Alexander was a Jewish inmate at Auschwitz who looked after 50 sets of Romani twins:

> "I remember one set of twins in particular: Guido and Ina, aged about four. One day, Mengele took them away. When they returned, they were in a terrible state: they had been sewn together, back to back, like Siamese twins. Their wounds were infected and oozing pus. They screamed day and night. Then their parents—I remember the mother's name was Stella—managed to get some morphine and they killed the children in order to end their suffering."

Development and execution
Origins

At 10 a.m. on April 1, 1933, members of the *Sturmabteilung* moved into place all over Germany, positioning themselves outside Jewish-owned businesses to deter customers. These stormtroopers are outside Israel's Department Store in Berlin. The signs read: "Germans! Defend yourselves! Don't buy from Jews." ("*Deutsche! Wehrt Euch! Kauft nicht bei Juden!*") The store was ransacked during Kristallnacht in 1938, then handed over to a non-Jewish family.

Yehuda Bauer, Raul Hilberg and Lucy Dawidowicz maintained that from the Middle Ages onward, German society and culture were suffused with anti-Semitism and there was a direct link from medieval pogroms to the Nazi death camps of the 1940s. Hans Küng has written that "Nazi anti-Judaism was the work of godless, anti-Christian criminals. But it would not have been possible without the almost two thousand years' pre-history of 'Christian' anti-Judaism . . ." The Nazi Party under Adolf Hitler came to power in Germany on January 30, 1933, and the persecution and exodus of Germany's 525,000 Jews began almost immediately. In *Mein Kampf* (1925), Hitler had been open about his hatred of Jews, and gave ample warning of his intention to drive them from Germany's political, intellectual, and cultural life. He did not write that he would attempt to exterminate them, but he is reported to have been more explicit in private. As early as 1922, he allegedly told Major Joseph Hell, at the time a journalist:

> "Once I really am in power, my first and foremost task will be the annihilation of the Jews. As soon as I have the power to do so, I will have gallows built in rows—at the Marienplatz in Munich, for example—as many as traffic allows. Then the Jews will be hanged indiscriminately, and they will remain hanging until they stink; they will hang there as long as the principles of hygiene permit. As soon as they have been untied, the next batch will be strung up, and so on down the line, until the last Jew in Munich has been exterminated. Other cities will follow suit, precisely in this fashion, until all Germany has been completely cleansed of Jews."

Legal repression and emigration

Throughout the 1930s, the legal, economic, and social rights of Jews were steadily restricted. In legally defining "who is Jew", the Nazis considered anyone of Jewish descent, even the descendents of converts who converted from Judaism after January 18, 1871, (the founding of the German Empire) were still considered Jews. Friedländer writes that, for the Nazis, Germany drew its strength for its "purity of blood" and its "rootedness in the sacred German earth." In 1933, a series of laws were passed which contained "Aryan paragraphs" to exclude Jews from key areas: the Law for the Restoration of the Professional Civil Service; the physicians' law; and the farm law, forbidding Jews from owning farms or taking part in agriculture. Jewish lawyers were disbarred, and in Dresden, Jewish lawyers and judges were dragged out of their offices and courtrooms, and beaten At the insistence of then president Hindenburg, Hitler added an exemption allowing Jewish civil servants who were veterans of the first world war, or whose fathers or sons had served, to remain in office. (Hindenburg was disturbed that people who had fought and bled for Germany would be forced from their state jobs.) Hitler revoked this exemption in 1937. Jews were excluded from schools and universities (the Law to prevent overcrowding in schools), from belonging to the Journalists' Association, and from being owners or editors of newspapers. The *Deutsche Allgemeine Zeitung* of April 27, 1933 wrote::

A self-respecting nation cannot, on a scale accepted up to now, leave its higher activities in the hands of people of racially foreign origin . . . Allowing the presence of too high a percentage of people of foreign origin in relation to their percentage in the general population could be interpreted as an acceptance of the superiority of other races, something decidedly to be rejected.

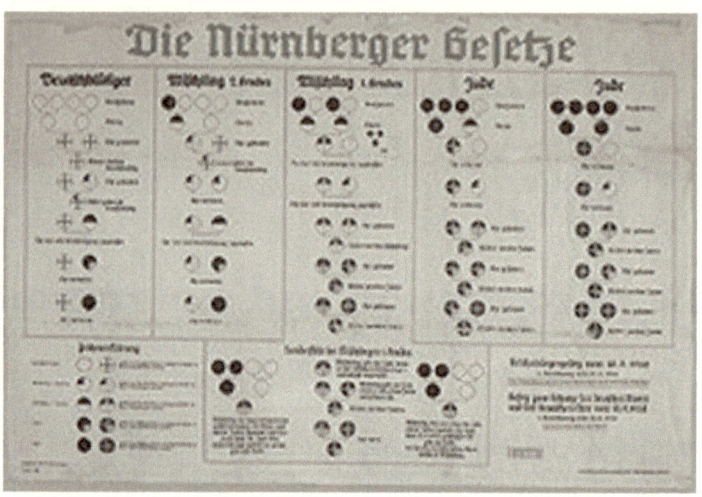

1935: Nazi definition of Jew, Mischling, and German and
legal consequences as per the Nuremberg Laws, simplified in a 1935 chart

In 1935, Hitler introduced the Nuremberg Laws, which: prohibited Jews from marrying Aryans, annulled existing marriages between Jews and Aryans (the Law for the protection of German blood and German honor), prohibited Jews from serving as civil servants, stripped German Jews of their citizenship and deprived them of all civil rights. In his speech introducing the laws, Hitler said that if the "Jewish problem" cannot be solved by these laws, it "must then be handed over by law to the National-Socialist Party for a final solution (*Endlösung*)." The expression "*Endlösung*" became the standard Nazi euphemism for the extermination of the Jews. In January 1939, he said in a public speech: "If international-finance Jewry inside and outside Europe should succeed once more in plunging the nations into yet another world war, the consequences will not be the Bolshevization of the earth and thereby the victory of Jewry, but the annihilation (*vernichtung*) of the Jewish race in Europe." Footage from this speech was used to conclude the 1940 Nazi propaganda movie The Eternal Jew (*Der ewige Jude*), whose purpose was to provide a rationale and blueprint for eliminating the Jews from Europe.

Jewish intellectuals were among the first to leave. The philosopher Walter Benjamin left for Paris on March 18, 1933. Novelist Leon Feuchtwanger went to Switzerland. The conductor Bruno Walter fled after being told that the hall of the Berlin Philharmonic would be burned down if he conducted a concert there: the *Frankfurter Zeitung* explained on April 6 that Walter and fellow conductor Otto Klemperer had been forced to flee because the government was unable to protect them against the "mood" of the German public, which had been provoked by "Jewish artistic liquidators." Albert Einstein was visiting the U.S. on January 30, 1933. He returned to Ostende in Belgium, never to set foot in Germany again, and calling events there a "psychic illness of the masses"; he was expelled from the Kaiser Wilhelm Society and the Prussian Academy of Sciences, and his citizenship was rescinded. When the Nazis annexed Austria in 1938, Sigmund Freud and his family fled from there to England. Saul Friedländer writes that when Max Liebermann, honorary president of the Prussian Academy of Arts, resigned his position, not one of his colleagues expressed a word of sympathy, and he died ostracized two years later. When the police arrived in 1943 with a stretcher to deport his 85-year-old bedridden widow, she committed suicide with an overdose of barbiturates rather than be taken.

Kristallnacht (1938)

Berlin's Fasanenstrasse synagogue after Kristallnacht,
November 9-10, 1938.

On November 7, 1938, Jewish minor Herschel Grünspan assassinated Nazi German diplomat Ernst vom Rath in Paris. This incident was used by the Nazis to initiate the transition from legal repression to large-scale outright violence against Jewish Germans. What the Nazis claimed to be spontaneous "public outrage" was mass pogroms conducted by the Nazi party and SA members and affiliates throughout Nazi Germany (then consisting of Germany proper, Austria and Sudetenland). The programs became known as *Reichskristallnacht* ("the Night of Broken Glass", literally *"Crystal Night"*), or *November pogroms*.[59] Jews were attacked and Jewish property was vandalized,[59] over 7,000 Jewish shops and 1,668 synagogues (almost every synagogue in Germany) were damaged or destroyed. The death toll is assumed to be much higher than the official number of 91 dead. 30,000[60] were sent to concentration camps, including Dachau, Sachsenhausen, Buchenwald, and Oranienburg concentration camp, where they were kept for several weeks, and released when they could either prove that they were about to emigrate in the near future, or transferred their property to the Nazis. The German Jewry was collectively made responsible for restitution of the material damage of the pogrom, amounting to several hundreds of thousand Reichsmark, and furthermore had to pay collectively an "atonement tax" of more than a billion Reichsmark.

After these pogroms, Jewish emigration from Nazi Germany accelerated, while public Jewish life in Germany ceased to exist.

Resettlement and deportation to colonies and reservations

Before the war, the Nazis considered mass exportation of German (and subsequently the European) Jewry from Europe. Plans to reclaim former German colonies such as Tanganyika and South West Africa for Jewish resettlement were halted by Adolf Hitler, who argued that no place where "so much blood of heroic Germans had been spilled" should be made available as a residence for the "worst enemies of the Germans". Diplomatic efforts were undertaken to convince the other former colonial powers, primarily the United Kingdom and France, to accept expelled Jews in their colonies. Areas considered for possible resettlement included British Palestine, Italian Abyssinia, British Guinea, British Rhodesia, French Madagascar, and Australia.

Of these areas, Madagascar was the most seriously discussed. Heydrich called the Madagascar Plan a "territorial final solution"; it was a remote location, and the island's unfavorable conditions would hasten deaths. In retrospect, although futile, this plan did constitute an important psychological step on the path to the Holocaust. Approved by Hitler in 1938, the resettlement planning was carried out by Eichmann's office. Once the mass killing of Jews began in 1941, however, resettlement planning was abandoned. The end of the Madagascar Plan was announced on February 10, 1942. The German Foreign Office was given the official explanation that, due to the war with the Soviet Union, Jews were to be "sent to the east".

Palestine was the only location to which any Nazi relocation plan succeeded in producing significant results, by means of an agreement begun in 1933 between the Zionist Federation of Germany (*die Zionistische Vereinigung für Deutschland*) and the Nazi government, the Haavara Agreement. This agreement resulted in the transfer of about 60,000 German Jews and $100 million from Germany to Palestine, up until the outbreak of World War II.

Early measures in German occupied Poland

Nazi Germany 1941, including areas annexed from Poland
and the General Government area.

The question of the treatment of the Jews became an urgent one for the Nazis after September 1939, when they invaded the western half of Poland, home to about two million Jews. The pre-war Second Polish Republic had been split between Nazi Germany and the Soviet Union, in the preceding Molotov-Ribbentrop Pact. Of the German share of Poland, the northwestern parts were annexed, while the southeastern parts were made the *Generalgouvernement* led by Hans Frank. The invasion led Britain, Australia, New Zealand, Canada, South Africa, and France to declare war—World War II had started.

Himmler's right-hand man, Reinhard Heydrich, recommended concentrating all the Polish Jews in ghettos in major cities, where they would be put to work for the German war industry. The ghettos would be in cities located on railway junctions, so that, in Heydrich's words, "future measures can be accomplished more easily." During his interrogation in 1961, Adolf Eichmann testified that the expression "future measures" was understood to mean "physical extermination."

German policemen tormenting a Jew in Rzeszów, Poland.

In September, Himmler appointed Reinhard Heydrich head of the Reich Main Security Office (*Reichssicherheitshauptamt* or RSHA, not be to confused with the RuSHA). This body was to oversee the work of the SS, the Security Police (SD), and the Gestapo in occupied Poland, and carry out the policy towards the Jews described in Heydrich's report. The first organized murders of Jews by German forces occurred during Operation Tannenberg and through Selbstschutz units. Later, the Jews were herded into ghettos, mostly in the General Government area of central Poland, where they were put to work under the Reich Labor Office headed by Fritz Saukel. Here many thousands were killed in various ways, and many more died of disease, starvation, and exhaustion, but there was still no program of systematic killing. There is no doubt, however, that the Nazis saw forced labor as a form of extermination. The expression *Vernichtung durch Arbeit* ("destruction through work") was frequently used.

Although it was clear by 1941 that the SS hierarchy, led by Himmler, was determined to embark on a policy of killing all the Jews under German control, there were important centers of opposition to this policy within the Nazi regime. The grounds for the opposition were mainly economic, not humanitarian. Hermann Göring, who had overall control of the German war industry, and the German army's Economics Department, representing the armaments industry, argued that the enormous Jewish labor force assembled in the General Government area (more than a million able-bodied workers) was an asset too valuable to waste while Germany was preparing to invade the Soviet Union.

Early measures in other occupied countries

When Nazi Germany occupied Norway, the Netherlands, Luxembourg, Belgium, and France in 1940, and Yugoslavia and Greece in 1941, anti-Semitic measures were also introduced into these countries, although the pace and severity varied greatly from country to country according to local political circumstances. Jews were removed from economic and cultural life and were subject to various restrictive laws, but physical deportation did not occur in most places before 1942. The Vichy regime in occupied France actively collaborated in persecuting French Jews. Germany's allies Italy, Hungary, Romania, Bulgaria and Finland were pressured to introduce ant-isemitic measures, but for the most part they did not comply until compelled to do so. The German puppet regime in Croatia, on the other hand, began actively persecuting Jews on its own initiative.

General Government and Lublin Reservation (Nisko Plan)

On September 28, 1939, Germany gained control over the Lublin area through the German-Soviet agreement in exchange for Lithuania. According to the Nisko Plan, they set up the Lublin-Lipowa Reservation in the area. The reservation was designated by Adolf Eichmann, who was assigned the task of removing all Jews from Germany, Austria and the Protectorate of Bohemia and Moravia. They shipped the first Jews to Lublin less than three weeks later on October 18, 1939. The first train loads consisted of Jews deported from Austria and the Protectorate of Bohemia and Moravia. By January 30, 1940, historians estimate a total of 78,000 Jews had been deported to Lublin from Germany, Austria and Czechoslovakia. On 12 and February 13, 1940, the Pomeranian Jews were deported to the Lublin reservation, resulting in Pomeranian Gauleiter Franz Schwede-Coburg to be the first to declare his Gau "*judenrein*" ("free of Jews"). On March 24, 1940 Hermann Göring put a hold on the Nisko Plan, and by the end of April, abandoned it entirely. By the time the Nisko Plan was stopped, the total number of Jews who had been transported to Nisko had reached 95,000, many of whom had died due to starvation.

In July 1940, due to the difficulties of supporting the increased population in the General Gouvernment, Hitler stopped deporting Jews there.[80] This was temporary, however, as military conditions made conquest of Britain doubtful. During 1940 and 1941, the murder of large numbers of Jews in German occupied Poland continued, and the deportation of Jews were deported to the General Gouvernment was undertaken. The deportation of Jews from Germany, particularly Berlin, was not officially completed until 1943. (Many Berlin Jews were able to survive in hiding.) by December 1939, 3.5 million Jews were crowded into the General Government area.

Concentration and labor camps
(1933-1945)

Major concentration and extermination camps: Auschwitz, Belzec, Bergen-Belsen, Chełmno, Dachau, Flossenbürg, Grini, Jasenovac, Klooga, Majdanek, Maly Trostinets, Mauthausen-Gusen, Ravensbrück, and Treblinka Nazi concentration camp badges: Black triangle, Pink triangle, Purple triangle, and Yellow badge

April 12, 1945: Lager Nordhausen, where 20,000 inmates
are believed to have died.

Leading up to the 1933 elections, the Nazis began intensifying acts of violence to wreak havoc among the opposition. With the cooperation of local authorities, they set up camps as concentration centers within Germany. One of the first was Dachau, which opened in March 1933. These early camps were meant to hold, torture, or kill only political prisoners, such as Communists and Social Democrats.

These early prisons—usually basements and storehouses—were eventually consolidated into full-blown, centrally run camps outside the cities. By 1942, six large extermination camps had been established in Nazi-occupied Poland. After 1939, the camps increasingly became places where Jews and POWs were either killed or forced to live as slave laborers, undernourished and tortured.[82] It is estimated that the Germans established 15,000 camps in the occupied countries, many of them in Poland.
New camps were focused on areas with large Jewish, Polish intelligentsia, communist, or Roma and Sinti populations, including inside Germany. The transportation of prisoners was often carried out under horrifying conditions using rail freight cars, in which many died before reaching their destination.

Extermination through labour, a means whereby camp inmates would literally be worked to death—or frequently worked until they could no longer perform work tasks, followed by their selection for extermination—was invoked as a further systematic extermination policy. Furthermore, while not designed as a method for systematic extermination, many camp prisoners died because of harsh overall conditions or from executions carried out on a whim after being allowed to live for days or months.

Upon admission, some camps tattooed prisoners with a prisoner ID. Those fit for work were dispatched for 12 to 14 hour shifts. Before and after, there were roll calls that could sometimes last for hours, with prisoners regularly dying of exposure.

Ghettos (1940-1945)

A child dying in the streets of the Warsaw Ghetto

After the invasion of Poland, the German Nazis established ghettos in which Jews and some Romani were confined, until they were eventually shipped to death camps to be murdered. The Warsaw Ghetto was the largest, with 380,000 people, and the Łódź Ghetto the second largest, holding 160,000. They were, in effect, immensely crowded prisons, described by Michael Berenbaum as instruments of "slow, passive murder." Though the Warsaw Ghetto contained 400,000 people—30% of the population of Warsaw—it occupied only 2.4% of the city's area, averaging 9.2 people per room.

From 1940 through 1942, starvation and disease, especially typhoid, killed hundreds of thousands. Over 43,000 residents of the Warsaw ghetto died there in 1941, more than one in ten; in Theresienstadt, more than half the residents died in 1942.

Each ghetto was run by a *Judenrat* (Jewish council) of German-appointed Jewish community leaders, who were responsible for the day-to-day running of the ghetto, including the provision of food, water, heat, medicine, and shelter, and who were also expected to make arrangements for deportations to extermination camps. Heinrich Himmler ordered the start of the deportations on July 19, 1942, and three days later, on July 22, the deportations from the Warsaw Ghetto began; over the next 52 days, until September 12, 300,000 people from Warsaw alone were transported in freight trains to the Treblinka extermination camp. Many other ghettos were completely depopulated.

Berenbaum writes that the defining moment that tested the courage and character of each *Judenrat* came when they were asked to provide a list of names of the next group to be deported. The *Judenrat* members went through the tried and tested methods of delay, bribery, stonewalling, pleading, and argumentation, until finally a decision had to be made. Some argued that their responsibility was to save the Jews who *could* be saved, and that therefore others had to be sacrificed; others argued, following Maimonides, that not a single individual should be handed over who had not committed a capital crime. *Judenrat* leaders such as Dr. Joseph Parnas in Lviv, who refused to compile a list, were shot. On October 14, 1942, the entire *Judenrat* of Byaroza committed suicide rather than cooperate with the deportations.

The first ghetto uprising occurred in September 1942 in the small town of Łachwa in southeast Poland. Though there were armed resistance attempts in the larger ghettos in 1943, such as the Warsaw Ghetto Uprising and the Białystok Ghetto Uprising, in every case they failed against the unmatched Nazi military force, and the remaining Jews were either killed or deported to the death camps, which the Germans euphemistically called "resettlement in the East."

Pogroms (1939-1942)

A number of deadly pogroms by local populations occurred during the Second World War, some with Nazi encouragement, and some spontaneously. This included the Naşi pogrom in Romania on June 30, 1941, in which as many 14,000 Jews were killed by Romanian residents and police, and the Jedwabne pogrom, in which between 380 and 1,600 Jews were killed by local Poles in July 1941.

Death squads (1941-1943)

Main articles: Einsatzgruppen and Mass graves in the Soviet Union

A member of Einsatzgruppe D is about to shoot a man sitting by a mass grave in Vinnitsa, Ukraine, in 1942. Present in the background are members of the German Army, the German Labor Service, and the Hitler Youth.[93] The back of the photograph is inscribed "The last Jew in Vinnitsa".

The German invasion of the Soviet Union in June 1941 opened a new phase. The Holocaust intensified after the Nazis occupied Lithuania, where close to 80 percent of Lithuanian Jews were exterminated before the end of the year. The Soviet territories occupied by early 1942, including all of Belarus, Estonia, Latvia, Lithuania, Ukraine, and Moldova and most Russian territory west of the line Leningrad-Moscow-Rostov, contained about three million Jews, including hundreds of thousands who had fled Poland in 1939. The remaining three million were left at the mercy of the Nazis.

Executions of Kiev Jews by German army mobile killing units (Einsatzgruppen) near Ivangorod in Ukraine. The photo was mailed from the Eastern Front to Germany and intercepted by a member of the Polish resistance.

Members of the local populations in certain occupied Soviet territories participated substantially in the killings of Jews and others.[96] In Lithuania, Latvia and western Ukraine, locals were deeply involved in the murder of Jews from the very beginning of the German occupation. The Latvian Arajs Kommando was an example of such an operation. To the south, Ukrainians killed approximately 24,000 Jews. In addition, Latvian and Lithuanian units left their own countries, and committed murders of Jews in Belarus, and Ukrainians served as concentration and death camp guards in Poland. Ustaše militia in Croatian areas also carried out acts of persecution.

Many of the mass killings were carried out in public, a change from previous practice. German witnesses to these killings emphasized the participation of the locals. Ultimately it was the Germans who organized and channelled the local participants in the Holocaust. The massacres committed by the *Einsatzgruppen* were usually justified under the grounds of anti-partisan or anti-bandit operations, but the German historian Andreas Hillgruber wrote that this was just a mere "excuse" for the German Army's considerable involvement in the Holocaust in Russia and the term war crimes and crimes against humanity were indeed correct labels for what happened. Hillgruber maintained that the slaughter of about 2.2 million defenceless men, women and children for the reasons of racist ideology cannot possibly be justified for any reason, and that those German generals who claimed that the *Einsatzgruppen* were a necessary anti-partisan response were lying.

Raul Hilberg writes that the German Einsatzgruppen commanders were ordinary citizens; the great majority were university-educated professionals. They used their skills to become efficient killers, according to Michael Berenbaum.

The large-scale killings of Jews in the occupied Soviet territories was assigned to SS formations called *Einsatzgruppen* ("task groups"), under the overall command of Heydrich. These had been used on a limited scale in Poland in 1939, but were now organized on a much larger scale. *Einsatzgruppe* A (commanded by SS-*Brigadeführer* Dr. Franz Stahlecker) was assigned to the Baltic area, *Einsatzgruppe* B (SS-*Brigadeführer* Artur Nebe) to Belarus, *Einsatzgruppe* C (SS-*Gruppenführer* Dr. Otto Rasch) to north and central Ukraine, and *Einsatzgruppe* D (SS-*Gruppenführer* Dr. Otto Ohlendorf) to Moldova, south Ukraine, the Crimea, and, during 1942, the north Caucasus. Of the four Einsatzgruppen, three were commanded by holders of doctorate degrees, of whom one (Rasch) held a double doctorate.

According to Ohlendorf at his trial, "the *Einsatzgruppen* had the mission to protect the rear of the troops by killing the Jews, Gypsies, Communist functionaries, active Communists, and all persons who would endanger the security." In practice, their victims were nearly all defenseless Jewish civilians (not a single *Einsatzgruppe* member was killed in action during these operations). By December 1941, the four *Einsatzgruppen* listed above had killed, respectively, 125,000, 45,000, 75,000, and 55,000 people—a total of 300,000 people—mainly by shooting or with hand grenades at mass killing sites outside the major towns.

The United States Holocaust Memorial Museum tells the story of one survivor of the Einsatzgruppen in Piryatin, Ukraine, when they killed 1,600 Jews on April 6, 1942, the second day of Passover:

I saw them do the killing. At 5:00 p.m. they gave the command, "Fill in the pits." Screams and groans were coming from the pits. Suddenly I saw my neighbor Ruderman rise from under the soil . . . His eyes were bloody and he was screaming: "Finish me off!" . . . A murdered woman lay at my feet. A boy of five years crawled out from under her body and began to scream desperately. "Mommy!" That was all I saw, since I fell unconscious.

The most notorious massacre of Jews in the Soviet Union was at a ravine called Babi Yar outside Kiev, where 33,771 Jews were killed in a single operation on September 29-30, 1941. The killing of all the Jews in Kiev was decided on by the military governor (Major-General Friedrich Eberhardt), the Police Commander for Army Group South (SS-*Obergruppenführer* Friedrich Jeckeln) and the *Einsatzgruppe* C Commander Otto Rasch. It was carried out by a mixture of SS, SD and Security Police, assisted by Ukrainian police.

On Monday the Jews of Kiev gathered by the cemetery, expecting to be loaded onto trains. The crowd was large enough that most of the men, women, and children could not have known what was happening until it was too late: by the time they heard the machine-gun fire, there was no chance to escape. All were driven down a corridor of soldiers, in groups of ten, and then shot. A truck driver described the scene:

One after the other, they had to remove their luggage, then their coats, shoes, and over garments and also underwear . . . Once undressed, they were led into the ravine which was about 150 meters long and 30 meters wide and a good 15 meters deep . . . When they reached the bottom of the ravine they were seized by members of the *Schutzpolizei* and made to lie down on top of Jews who had already been shot . . . The corpses were literally in layers. A police marksman came along and shot each Jew in the neck with a submachine gun . . . I saw these marksmen stand on layers of corpses and shoot one after the other . . . The marksman would walk across the bodies of the executed Jews to the next Jew, who had meanwhile lain down, and shoot him.

From left to right; Heinrich Himmler, Reinhard Heydrich, and Karl Wolff (second from the right) at the Obersalzberg, May 1939. Wolff wrote in his diary that Himmler had vomited after witnessing the mass shooting of 100 Jews.

In August 1941 Himmler travelled to Minsk, where he personally witnessed 100 Jews being shot in a ditch outside the town, an event described by SS-*Obergruppenführer* Karl Wolff in his diary. "Himmler's face was green. He took out his handkerchief and wiped his cheek where a piece of brain had squirted up on to it. Then he vomited." After recovering his composure, he lectured the SS men on the need to follow the "highest moral law of the Party" in carrying out their tasks.

New methods of mass murder

Starting in December 1939, the Nazis introduced new methods of mass murder by using gas. First experimental vans, equipped with gas cylinders and a sealed trunk compartment, were used to kill mental care clients of sanatoria in Pomerania, East Prussia, and occupied Poland, as part of an operation termed Aktion T4. In the Sachsenhausen concentration camp, larger vans holding up to 100 people were used in a similar way since November 1941, yet the gas did not come from a cylinder but directly from the engine's exhaust. These vans were introduced to the Chełmno concentration camp in December 1941, and another 15 of them were used by the death squads in the occupied Soviet Union. These gas vans were developed and run under supervision of the *Reichssicherheitshauptamt* (Reich Main Security Bureau), and were used to kill about 500,000 people, primarily Jews, but also Romani and others. The vans were carefully monitored and after a month of observation a report stated that 'ninety seven thousand have been processed using three vans, without any defects showing up in the machines'.

A need for new mass murder techniques was also expressed by Hans Frank, governor of the General Government, who noted that this many people could not be simply shot. "We shall have to take steps, however, designed in some way to eliminate them." It was this dilemma which led the SS to experiment with large-scale killings using poison gas. Finally, SS *Obersturmführer* Christian Wirth seems to have been the inventor of the gas chamber.

Wannsee Conference and the Final Solution (1942-1945)

The dining room of the Wannsee villa, where the Wannsee conference took place. The 15 men seated at the table on January 20, 1942 to discuss the "final solution of the Jewish question" were considered the best and the brightest in the Reich.

Facsimiles of the minutes of the Wannsee Conference. This page lists the number of Jews in every European country.

Auschwitz I

The railway line leading to the death camp at Auschwitz II (Birkenau).

Empty poison gas canisters used to kill inmates and piles of hair shaven from their heads are stored in the museum at Auschwitz II.

The ruins of the Crematorium II gas chamber at Auschwitz II (Birkenau).
Holocaust scholar Robert Jan van Pelt comments that more people lost their
lives in this room than in any other room on Earth: 500,000 people.

*Those present at the conference: Josef Bühler, Adolf Eichmann, Roland Freisler,
Reinhard Heydrich, Otto Hofmann, Gerhard Klopfer, Friedrich Wilhelm
Kritzinger, Rudolf Lange, Georg Leibbrandt, Martin Luther, Heinrich Müller,
Erich Neumann, Karl Eberhard Schöngarth, Wilhelm Stuckart*

By the end of 1941, Himmler was becoming increasingly impatient with
the progress of the Final Solution. His main opponent was Göring, who
had succeeded in exempting Jewish industrial workers from the orders to
deport all Jews to the General Government and who had allied himself
with the Army commanders who were opposing the extermination of the
Jews out of a mixture of economic calculation, distaste for the SS and
humanitarian sentiment. Although Göring's power had declined since the
defeat of his Luftwaffe in the Battle of Britain, he still had privileged access
to Hitler.

The Nazis methodically tracked the progress of the Holocaust in thousands
of reports and documents. Pictured is the Höfle Telegram sent to Adolf
Eichmann in January, 1943, that reported that 1,274,166 Jews had been
killed in the four Aktion Reinhard camps during 1942.

Heydrich therefore convened the Wannsee Conference on January 20, 1942 at a villa, *Am Großen Wannsee* No. 56-58, in the suburbs of Berlin to finalize a plan for the extermination of the Jews. The plan became known (after Heydrich) as *Aktion Reinhard* (Operation Reinhard). Present were Heydrich, Eichmann, Heinrich Müller (head of the Gestapo), and representatives of the Ministry for the Occupied Eastern Territories, the Ministry for the Interior, the Four Year Plan Office, the Ministry of Justice, the General Government in Poland (where over two million Jews still lived), the Foreign Office, the Race and Resettlement Office, and the Nazi Party, and the office responsible for distributing Jewish property. Also present was SS-*Sturmbannführer* Rudolf Lange, the SD commander in Riga, who, with Friedrich Jeckeln had recently carried out the liquidation of 24,000 Latvian Jews from the Riga ghetto in the Rumbula massacre.

Michael Berenbaum writes that the 15 men seated at the table were considered the best and the brightest; more than half of them held doctorates from German universities.

A plan was presented for killing all the Jews in Europe, including 330,000 Jews in England and 4,000 in Ireland, although the minutes taken by Eichmann refer to this only through euphemisms, such as " . . . emigration has now been replaced by evacuation to the East. This operation should be regarded only as a provisional option, though in view of the coming final solution of the Jewish question it is already supplying practical experience of vital importance."

The officials were told there were 2.3 million Jews in the General Government, 850,000 in Hungary, 1.1 million in the other occupied countries, and up to 5 million in the Soviet Union (although only 3 million of these were in areas under German occupation)—a total of about 6.5 million. These would all be transported by train to extermination camps (*Vernichtungslager*) in Poland, where those unfit for work would be gassed at once. In some camps, such as Auschwitz, those fit for work would be kept alive for a while, but eventually all would be killed. Göring's representative, Dr. Erich Neumann, gained a limited exemption for some classes of industrial workers.

During 1942, in addition to Auschwitz, five other camps were designated as extermination camps (*Vernichtungslager*) for the carrying out of the

Reinhard plan.[127][128] Two of these, Chełmno (also known as Kulmhof) and Majdanek were already functioning as labor camps: these now had extermination facilities added to them. Three new camps were built for the sole purpose of killing large numbers of Jews as quickly as possible, at Belzec, Sobibór and Treblinka. A seventh camp, at Maly Trostinets in Belarus, was also used for this purpose. Jasenovac was an extermination camp where mostly ethnic Serbs were killed.

Extermination camps are frequently confused with concentration camps such as Dachau and Belsen, which were mostly located in Germany and intended as places of incarceration and forced labor for a variety of enemies of the Nazi regime (such as Communists and homosexuals). They should also be distinguished from slave labor camps, which were set up in all German-occupied countries to exploit the labor of prisoners of various kinds, including prisoners of war. In all Nazi camps there were very high death rates as a result of starvation, disease and exhaustion, but only the extermination camps were designed specifically for mass killing.

The extermination camps were run by SS officers, but most of the guards were Ukrainian or Baltic auxiliaries. Regular German soldiers were kept well away.

Gas chambers

At the extermination camps with gas chambers all the prisoners arrived by train. Sometimes entire trainloads were sent straight to the gas chambers, but usually the camp doctor on duty subjected individuals to selections, where a small percentage were deemed fit to work in the slave labor camps; the majority were taken directly from the platforms to a reception area where all their clothes and other possessions were seized by the Nazis to help fund the war. They were then herded naked into the gas chambers. Usually they were told these were showers or delousing chambers, and there were signs outside saying "baths" and "sauna." They were sometimes given a small piece of soap and a towel so as to avoid panic, and were told to remember where they had put their belongings for the same reason. When they asked for water because they were thirsty after the long journey in the cattle trains, they were told to hurry up, because coffee was waiting for them in the camp, and it was getting cold.

According to Rudolf Höß, commandant of Auschwitz, bunker 1 held 800 people, and bunker 2 held 1,200. Once the chamber was full, the doors were screwed shut and solid pellets of Zyklon-B were dropped into the chambers through vents in the side walls, releasing toxic HCN, or hydrogen cyanide. Those inside died within 20 minutes; the speed of death depended on how close the inmate was standing to a gas vent, according to Höß, who estimated that about one third of the victims died immediately. Joann Kremer, an SS doctor who oversaw the gassings, testified that: "Shouting and screaming of the victims could be heard through the opening and it was clear that they fought for their lives." When they were removed, if the chamber had been very congested, as they often were, the victims were found half-squatting, their skin colored pink with red and green spots, some foaming at the mouth or bleeding from the ears.

The gas was then pumped out, the bodies were removed (which would take up to four hours), gold fillings in their teeth were extracted with pliers by dentist prisoners, and women's hair was cut. The floor of the gas chamber was cleaned, and the walls whitewashed. The work was done by the *Sonderkommando* prisoners, Jews who hoped to buy themselves a few extra months of life. In crematoria 1 and 2, the *Sonderkommando* lived in an attic above the crematoria; in crematoria 3 and 4, they lived inside the gas chambers. When the *Sonderkommando* had finished with the bodies, the SS conducted spot checks to make sure all the gold had been removed from the victims' mouths. If a check revealed that gold had been missed, the *Sonderkommando* prisoner responsible was thrown into the furnace alive as punishment.

At first, the bodies were buried in deep pits and covered with lime, but between September and November 1942, on the orders of Himmler, they were dug up and burned. In the spring of 1943, new gas chambers and crematoria were built to accommodate the numbers.

Another improvement we made over Treblinka was that we built our gas chambers to accommodate 2,000 people at one time, whereas at Treblinka their 10 gas chambers only accommodated 200 people each. The way we selected our victims was as follows: we had two SS doctors on duty at Auschwitz to examine the incoming transports of prisoners. The prisoners would be marched by one of the doctors who would make spot decisions as they walked by. Those who were fit for work were sent into the Camp. Others were sent immediately to the extermination plants. Children of tender years were invariably exterminated, since by reason of their youth they were unable to work. Still another improvement we made over Treblinka was that at Treblinka the victims almost always knew that they were to be exterminated and at Auschwitz we endeavored to fool the victims into thinking that they were to go through a delousing process. Of course, frequently they realized our true intentions and we sometimes had riots and difficulties due to that fact. Very frequently women would hide their children under the clothes but of course when we found them we would send the children in to be exterminated. We were required to carry out these exterminations in secrecy but of course the foul and nauseating stench from the continuous burning of bodies permeated the entire area and all of the people living in the surrounding communities knew that exterminations were going on at Auschwitz.

Jewish resistance

Jews captured and forcibly pulled out from dugouts by the Germans during the Warsaw Ghetto uprising. The photo is from Jurgen Stroop's report to Heinrich Himmler.

Warsaw Ghetto uprising

Yehuda Bauer and other historians argue that resistance consisted not only of physical opposition, but of any activity that gave the Jews dignity and humanity in humiliating and inhumane conditions.

In every ghetto, in every deportation train, in every labor camp, even in the death camps, the will to resist was strong, and took many forms. Fighting with the few weapons that would be found, individual acts of defiance and protest, the courage of obtaining food and water under the threat of death, the superiority of refusing to allow the Germans their final wish to gloat over panic and despair. Even passivity was a form of resistance. To die with dignity was a form of resistance. To resist the demoralizing, brutalizing force of evil, to refuse to be reduced to the level of animals, to live through the torment, to outlive the tormentors, these too were acts of resistance. Merely to give a witness of these events in testimony was, in the end, a contribution to victory. Simply to survive was a victory of the human spirit."

The Jewish Tragedy

There are many examples of Jewish resistance, most notably the Warsaw Ghetto Uprising of January 1943, when thousands of poorly armed Jewish fighters held the SS at bay for four weeks before being crushed by overwhelmingly superior forces. According to Jewish accounts, several hundred Germans were killed, while the Germans claimed to have lost 17 dead and 93 wounded. 13,000 Jews were killed during the uprising, and 57,885 were deported and gassed according to German figures. This uprising was followed by the uprising in the Treblinka extermination camp in May 1943, when about 200 inmates escaped from the camp after overpowering the guards. They killed a number of German guards and set the camp buildings ablaze, but 900 inmates were also killed, and out of the 600 who successfully escaped, only 40 survived the war. Two weeks later, there was an uprising in the Białystok ghetto. In September, there was a short-lived uprising in the Vilnius ghetto. In October, 600 Jewish prisoners, including Jewish Soviet prisoners of war, attempted an escape at the Sobibór death camp. The prisoners killed 11 German SS officers and a number of camp guards. However, the killings were discovered, and the inmates were forced to run for their lives under heavy fire. 300 of the prisoners were killed during the escape. Most of the survivors either died in the minefields surrounding the camp or were recaptured and executed. About 60 survived and joined the Soviet partisans. On October 7, 1944, 250 Jewish *Sonderkommandos* (laborers) at Auschwitz attacked their guards and blew up Crematorium IV with explosives female prisoners had smuggled in from a nearby factory. Three German guards were killed during the uprising, one of whom was stuffed into an oven. The Sonderkommandos attempted a mass breakout, but all 250 were killed soon after.

An estimated 20,000 to 30,000 Jewish partisans (see the list at the top of this section) actively fought the Nazis and their collaborators in Eastern Europe.[141] They engaged in guerilla warfare and sabotage against the Nazis, instigated Ghetto uprisings, and freed prisoners. In Lithuania alone, they killed approximately 3,000 German soldiers. As many as 1.4 million Jewish soldiers fought in the Allied armies. Of these, approximately 40% served in the Red Army.[142] The Jewish Brigade, a unit of 5,000 Jewish volunteers from the British Mandate of Palestine fought in the British Army. German-speaking Jewish volunteers from the Special Interrogation Group performed commando and sabotage operations against the Nazis behind front lines in the Western Desert Campaign.

In occupied Poland and Soviet territories, thousands of Jews fled into the swamps or forests and joined the partisans, although the partisan movements did not always welcome them. In Lithuania and Belarus, an area with a heavy concentration of Jews, and also an area which suited partisan operations, Jewish partisan groups saved thousands of Jewish civilians from extermination. No such opportunities existed for the Jewish populations of cities such as Budapest. However in Amsterdam, and other parts of the Netherlands, many Jews were active in the Dutch Resistance. Joining the partisans was an option only for the young and the fit who were willing to leave their families. Many Jewish families preferred to die together rather than be separated.

French Jews were also highly active in the French Resistance, which conducted a massive guerilla campaign against the Nazis and Vichy French authorities, assisted the Allies in their sweep across France, and participated in many liberations of occupied French cities. Although Jews made up only one percent of the French population, they made up fifteen to twenty percent of the French Resistance. The Jewish youth movement EEIF, which had originally shown support for the Vichy regime, was banned in 1943, and many of its older members formed armed resistance units. Zionist Jews also formed the Armee Juive (Jewish Army), which participated in armed resistance under a Zionist flag, and smuggled Jews out of the country. Both organizations merged in 1944, and participated in the liberations of Paris, Lyon, Toulouse, Grenoble, and Nice.

"Many people think the Jews went to their deaths like sheep to the slaughter, and that's not true—it's absolutely not true. I worked closely with many

Jewish people in the Resistance, and I can tell you, they took much greater risks than I did."

For the great majority of Jews resistance could take only the passive forms of delay, evasion, negotiation, bargaining and, where possible, bribery of German officials. The Nazis encouraged this by forcing the Jewish communities to police themselves, through bodies such as the Reich Association of Jews (*Reichsvereinigung der Juden*) in Germany and the Jewish Councils (*Judenrate*) in the urban ghettos in occupied Poland. They held out the promise of concessions in exchange for each surrender, enmeshing the Jewish leadership so deeply in well-intentioned compromise that a decision to stand and fight was never possible. Holocaust survivor Alexander Kimel wrote: "The youth in the Ghettos dreamed about fighting. I believe that although there were many factors that inhibited our responses, the most important factors were isolation and historical conditioning to accepting martyrdom."

The historical conditioning of the Jewish communities of Europe to accept persecution and avert disaster through compromise and negotiation was the most important factor in the failure to resist until the very end. The Warsaw Ghetto uprising took place only when the Jewish population had been reduced from 500,000 to 100,000, and it was obvious that no further compromise was possible. Paul Johnson writes: "The Jews had been persecuted for a millennium and a half and had learned from long experience that resistance cost lives rather than saved them. Their history, their theology, their folklore, their social structure, even their vocabulary trained them to negotiate, to pay, to plead, to protest, not to fight."

The Jewish communities were also systematically deceived about German intentions, and were cut off from most sources of news from the outside world. The Germans told the Jews that they were being deported to work camps—euphemistically calling it "resettlement in the East"—and maintained this illusion through elaborate deceptions all the way to the gas chamber doors (which were marked with labels stating that the chambers were for removal of lice) to avoid uprisings. As photographs testify, Jews disembarked at the railway stations at Auschwitz and other extermination camps carrying sacks and suitcases, clearly having no idea of the fate that awaited them. Rumours of the reality of the extermination camps filtered back only slowly to the ghettos, and were usually not believed, just as they were not believed when couriers such as Jan Karski, the Polish resistance fighter, conveyed them to the western Allies.

Climax

Heydrich was assassinated in Prague in June 1942. He was succeeded as head of the RSHA by Ernst Kaltenbrunner. Kaltenbrunner and Eichmann, under Himmler's close supervision, oversaw the climax of the Final Solution. During 1943 and 1944, the extermination camps worked at a furious rate to kill the hundreds of thousands of people shipped to them by rail from almost every country within the German sphere of influence. By the spring of 1944, up to 8,000 people were being gassed every day at Auschwitz.

Despite the high productivity of the war industries based in the Jewish ghettos in the General Government, during 1943 they were liquidated, and their populations shipped to the camps for extermination. The largest of these operations, the deportation of 100,000 people from the Warsaw Ghetto in early 1943, provoked the Warsaw Ghetto Uprising, which was suppressed with great brutality. At the same time, rail shipments arrived regularly from western and southern Europe. Few Jews were shipped from the occupied Soviet territories to the camps: the killing of Jews in this zone was left in the hands of the SS, aided by locally recruited auxiliaries. In any case, by the end of 1943 the Germans had been driven from most Soviet territory.

Shipments of Jews to the camps had priority on the German railways, and continued even in the face of the increasingly dire military situation after the Battle of Stalingrad at the end of 1942 and the escalating Allied air attacks on German industry and transport. Army leaders and economic managers complained at this diversion of resources and at the killing of irreplaceable skilled Jewish workers. By 1944, moreover, it was evident to most Germans not blinded by Nazi fanaticism that Germany was losing the war. Many senior officials began to fear the retribution that might await Germany and them personally for the crimes being committed in their name. But the power of Himmler and the SS within the German Reich was too great to resist, and Himmler could always evoke Hitler's authority for his demands.

Budapest, Hungary—Captured Jewish women in
Wesselényi Street, 20-22 October 1944

Budapest, Hungary—Hungarian and German soldiers
drive arrested Jews into the municipal theatre dated October 1944.

In October 1943, Himmler gave a speech to senior Nazi Party officials gathered in Posen (Poznań in western Poland). Here he came closer than ever before to stating explicitly that he was intent on exterminating the Jews of Europe:

I may here in this closest of circles allude to a question which you, my party comrades, have all taken for granted, but which has become for me the most difficult question of my life, the Jewish question . . . I ask of you that what I say in this circle you really only hear and never speak of . . . We come to the question: how is it with the women and children? I have resolved even here on a completely clear solution. I do not consider myself justified in eradicating the men—so to speak killing them or ordering them to be killed—and allowing the avengers in the shape of the children to grow up . . . The difficult decision had to be taken, to cause this people to disappear from the earth.

The audience for this speech included Admiral Karl Dönitz and Armaments Minister Albert Speer, both of whom successfully claimed at the Nuremberg trials that they had had no knowledge of the Final Solution. The text of this speech was not known at the time of their trials.

The scale of extermination slackened somewhat at the beginning of 1944 once the ghettos in occupied Poland were emptied, but in March 19, 1944, Hitler ordered the military occupation of Hungary, and Eichmann was dispatched to Budapest to supervise the deportation of Hungary's 800,000 Jews. Hitler had personally complained to the Hungarian regent Admiral Miklós Horthy on the previous day, March 18, 1944, that:

> *"Hungary did nothing in the matter of the Jewish problem, and was not prepared to settle accounts with the large Jewish population in Hungary."*

More than half of them were shipped to Auschwitz in the course of the year. The commandant, Rudolf Höß, said at his trial that he killed 400,000 Hungarian Jews in three months. This operation met strong opposition within the Nazi hierarchy, and there were some suggestions that Hitler should offer the Allies a deal under which the Hungarian Jews would be spared in exchange for a favorable peace settlement. There were unofficial negotiations in Istanbul between Himmler's agents, British agents, and representatives of Jewish organizations, and at one point an attempt by Eichmann to exchange one million Jews for 10,000 trucks—the so-called "blood for goods" proposal—but there was no real possibility of such a deal being struck.

Escapes, publication of news of the death camps
(April-June 1944)

Bratislava, June-July 1944. Rudolf Vrba (right) escaped from Auschwitz on April 7, 1944, bringing the first credible news to the world of the mass murder that was taking place there. Arnost Rosin (left), escaped on May 27, 1944.

Escapes from the camps were few, but not unknown. The few Auschwitz escapes that succeeded were made possible by the Polish underground inside the camp and local people outside. In 1940, the Auschwitz commandant reported that "the local population is fanatically Polish and . . . prepared to take any action against the hated SS camp personnel. Every prisoner who managed to escape can count on help the moment he reaches the wall of a first Polish farmstead."

In February 1942, an escaped inmate from the Chełmno extermination camp, Jacob Grojanowski, reached the Warsaw Ghetto, where he gave detailed information about the Chełmno camp to the Oneg Shabbat group. His report, which became known as the Grojanowski Report, was smuggled out of the ghetto through the channels of the Polish underground to the Delegatura, and reached London by June 1942. It is unclear what was done with the report at that point. In the meantime, by the 1ˢᵗ of February, the United States Office of War Information had decided not to release information about the extermination of the Jews because it was felt that it would mislead the public into thinking the war was simply a Jewish problem.

In December 1942, the western Allies released a declaration, publicized on the *New York Times* front page, that described how "Hitler's oft-repeated intention to exterminate the Jewish people in Europe" was being carried out and which declared that they "condemn in the strongest possible terms this bestial policy of cold-blooded extermination."

Jan Karski, 1944.

In 1942, Jan Karski reported to the Polish, British and U.S. governments on the situation in Poland, especially the destruction of the Warsaw Ghetto and the Holocaust of the Jews. He met with Polish politicians in exile including the prime minister, as well as members of political parties such as the PPS, SN, SP, SL, Jewish Bund and Poalei Zion. He also spoke to Anthony Eden, the British foreign secretary, and included a detailed statement on what he had seen in Warsaw and Bełżec. In 1943 in London he met the then-well-known journalist Arthur Koestler. He then traveled to the United States and reported to President Franklin D. Roosevelt. His report was a major factor in informing the West.

In July 1943, Karski again personally reported to Roosevelt about the situation in Poland. During their meeting Roosevelt suddenly interrupted his report and asked about the condition of horses in occupied Poland. He also met with many other government and civic leaders in the United States, including Felix Frankfurter, Cordell Hull, William Joseph Donovan, and Stephen Wise. Karski also presented his report to media, bishops of various denominations (including Cardinal Samuel Stritch), members of the Hollywood film industry and artists, but without success. Many of those he spoke to did not believe him, or supposed that his testimony was much exaggerated or was propaganda from the Polish government in exile.

In 1943, the news about gassing Jews was broadcast from London to The Netherlands. It was also published in illegal newspapers of the Dutch resistance, like in the issue of Het Parool of September 27, 1943. However, the news was so unbelievable that many assumed it was merely war propaganda. The publications were halted because they were counter-productive for the Dutch resistance. Nevertheless, many Jews were warned that they would be murdered, but as escape was impossible for most of them, they preferred to believe that the warnings were false.

Auschwitz concentration cp photos of Pilecki (1941)

In September 1940, Captain Witold Pilecki, a member of the Polish underground and a soldier of the Home Army, worked out a plan to enter Auschwitz and volunteered to be sent there, the only known person to volunteer to be imprisoned at Auschwitz. He organized an underground network Związek Organizacji Wojskowej (*translation: "Union of Military Organizations"*) that was ready to initiate an uprising but it was decided that the probability of success was too low for the uprising to succeed. UMO's numerous and detailed reports became later a principal source of intelligence on Auschwitz for the Western Allies. Pilecki escaped from Auschwitz with information that became the basis of a two-part report in August 1943 that was sent to the Office of Strategic Services (OSS) in London. The report included details about the gas chambers, about "selection", and about the sterilization experiments. It stated that there were three crematoria in Birkenau able to burn 10,000 people daily, and that 30,000 people had been gassed in one day. The author wrote: "History knows no parallel of such destruction of human life." Raul Hilberg writes that the report was filed away with a note that there was no indication as to the reliability of the source. When Pilecki returned to Poland after the war the communist authorities arrested and accused him of spying for the Polish government in exile. He was sentenced to death in a show trial and was executed on May 25, 1948.

Before Pilecki escaped from Auschwitz the most spectacular escape took place on 20 June 1942, when Ukrainian Eugeniusz Bendera and three Poles, Kazimierz Piechowski, Stanisław Gustaw Jaster and Józef Lempart made a daring escape.[169] The escapees were dressed as members of the SS-Totenkopfverbände, fully armed and in an SS staff car. They drove out the main gate in a stolen Rudolf Hoss automobile Steyr 220 with a smuggled first report from Witold Pilecki to Polish resistance about the Holocaust. The Germans never recaptured any of them.

Rudolf Vrba and Alfred Wetzler, Jewish inmates, escaped from Auschwitz in April 1944, eventually reaching Slovakia. The 32-page document they dictated to Jewish officials about the mass murder at Auschwitz became known as the Vrba-Wetzler report. Vrba had an eidetic memory and had worked on the *Judenrampe*, where Jews disembarked from the trains to be "selected" either for the gas chamber or slave labor. The level of detail with which he described the transports allowed Slovakian officials to compare his account with their own deportation records, and the corroboration convinced the Allies to take the report seriously.

Two other Auschwitz inmates, Arnost Rosin and Czesław Mordowicz escaped on May 27, 1944, arriving in Slovakia on June 6, the day of the Normandy landing (D-Day). Hearing about Normandy, they believed the war was over and got drunk to celebrate, using dollars they'd smuggled out of the camp. They were arrested for violating currency laws, and spent eight days in prison, before the *Judenrat* paid their fines. The additional information they offered the Judenrat was added to Vrba and Wetzler's report and became known as the Auschwitz Protocols. They reported that, between May 15 and May 27, 1944, 100,000 Hungarian Jews had arrived at Birkenau, and had been killed at an unprecedented rate, with human fat being used to accelerate the burning.

The subsequent pressure from world leaders persuaded Miklós Horthy to bring the mass deportations of Jews from Hungary to Auschwitz to a halt on July 9, saving up to 200,000 Jews from the extermination camps.

On November 14, 2001, in the 150th anniversary issue, The New York Times ran an article by former editor Max Frankel reporting that before and during World War II, the Times had maintained a strict policy in their news reporting and editorials to minimize reports on the Holocaust. The Times accepted the detailed analysis and findings of journalism professor Laurel Leff, who had published an article the year before in the Harvard International Journal of the Press and Politics, that the New York Times had deliberately suppressed news of the Third Reich's persecution and murder of Jews. Leff concluded that New York Times reporting and editorial polices made it virtually impossible for American Jews to impress Congress, church or government leaders with the importance of helping Europe's Jews.

Death marches (1944-1945)

Children from Auschwitz liberated by the Red Army in January, 1945. Although most children were immediately killed upon arrival, this group includes Jewish twins kept alive to be used in Mengele's medical experiments

By mid 1944, the Final Solution had largely run its course. Those Jewish communities within easy reach of the Nazi regime had been largely exterminated, in proportions ranging from more than 90 percent in Poland to about 25 percent in France. In May, Himmler claimed in a speech that "The Jewish question in Germany and the occupied countries has been solved." During 1944, in any case, the task became steadily more difficult.

German armies were evicted from the Soviet Union, the Balkans and Italy, and German allies were either defeated or were switching sides to the Allies. In June, the western Allies landed in France. Allied air attacks and the operations of partisans made rail transport increasingly difficult and the objections of the military to the diversion of rail transport for carrying Jews to Poland more urgent and harder to ignore.

At this time, as the Soviet armed forces approached, the camps in eastern Poland were closed down, any surviving inmates being shipped west to camps closer to Germany, first to Auschwitz and later to Gross Rosen in Silesia. Auschwitz itself was closed as the Soviets advanced through Poland. The last 13 prisoners, all women, were killed in Auschwitz II on November 25, 1944; records show they were "*unmittelbar getötet*" ("killed outright"), leaving open whether they were gassed or otherwise disposed of.

Despite the desperate military situation, great efforts were made to conceal evidence of what had happened in the camps. The gas chambers were dismantled, the crematoria dynamited, mass graves dug up and the corpses cremated, and Polish farmers were induced to plant crops on the sites to give the impression that they had never existed. In October 1944, Himmler, who is believed to have been negotiating a secret deal with the Allies behind Hitler's back, ordered an end to the Final Solution. But the hatred of the Jews in the ranks of the SS was so strong that Himmler's order was generally ignored. Local commanders continued to kill Jews, and to shuttle them from camp to camp by forced "death marches" until the last weeks of the war.

Already sick after months or years of violence and starvation, prisoners were forced to march for tens of miles in the snow to train stations; then transported for days at a time without food or shelter in freight trains with open carriages; and forced to march again at the other end to the new camp. Those who lagged behind or fell were shot. Around 250,000 Jews died during these marches.

The largest and best-known of the death marches took place in January 1945, when the Soviet army advanced on Poland. Nine days before the Soviets arrived at Auschwitz, the SS marched 60,000 prisoners out of the camp toward Wodzislaw, 56 km (35 miles) away, where they were put on freight trains to other camps. Around 15,000 died on the way. Elie Wiesel and his father, Shlomo, were among the marchers:

An icy wind blew in violent gusts. But we marched without faltering. Pitch darkness. Every now and then, an explosion in the night. They had orders to fire on any who could not keep up. Their fingers on the triggers, they did not deprive themselves of this pleasure. If one of us had stopped for a second, a sharp shot finished off another filthy son of a bitch. Near me, men were collapsing in the dirty snow. Shots.

Liberation

A grave inside Bergen-Belsen

The first major camp, Majdanek, was discovered by the advancing Soviets on July 23, 1944. Auschwitz was liberated, also by the Soviets, on January 27, 1945; Buchenwald by the Americans on April 11; Bergen-Belsen by the British on April 15; Dachau by the Americans on April 29; Ravensbrück by the Soviets on the same day; Mauthausen by the Americans on May 5; and Theresienstadt by the Soviets on May 8. Treblinka, Sobibor, and Belzec were never liberated, but were destroyed by the Nazis in 1943. Colonel William W. Quinn of the U.S. 7th Army said of Dachau: "There our troops found sights, sounds, and stenches horrible beyond belief, cruelties so enormous as to be incomprehensible to the normal mind."

In most of the camps discovered by the Soviets, almost all the prisoners had already been removed, leaving only a few thousand alive—7,000 inmates were found in Auschwitz, including 180 children who had been experimented on by doctors. Some 60,000 prisoners were discovered at Bergen-Belsen by the British 11th Armoured Division, 13,000 corpses lay unburied, and another 10,000 died from typhus or malnutrition over the following weeks. The British forced the remaining SS guards to gather up the corpses and place them in mass graves.

Here over an acre of ground lay dead and dying people. You could not see which was which . . . The living lay with their heads against the corpses and around them moved the awful, ghostly procession of emaciated, aimless people, with nothing to do and with no hope of life, unable to move out of your way, unable to look at the terrible sights around them . . . Babies had been born here, tiny wizened things that could not live . . . A mother, driven mad, screamed at a British sentry to give her milk for her child, and thrust the tiny mite into his arms . . . He opened the bundle and found the baby had been dead for days. This day at Belsen was the most horrible of my life.

Victims and death toll

Members of the *Sonderkommando* burn corpses in the fire pits at Auschwitz II-Birkenau. Courtesy of the Auschwitz-Birkenau museum, Poland.

The number of victims depends on which definition of "the Holocaust" is used. Donald Niewyk and Francis Nicosia write in *The Columbia Guide to the Holocaust* that the term is commonly defined as the mass murder, and attempt to wipe out, European Jewry, which would bring the total number of victims to just under six million—around 78 percent of the 7.3 million Jews in occupied Europe at the time.

Broader definitions include approximately 2 to 3 million Soviet POWs, 2 million ethnic Poles, up to 1,500,000 Romani, 200,000 handicapped, political and religious dissenters, 15,000 homosexuals and 5,000 Jehovah's Witnesses, bringing the death toll to around 11 million. The broadest definition would include 6 million Soviet civilians, raising the death toll to 17 million. R.J. Rummel estimates the total democide death toll of Nazi Germany to be 21 million. Other estimates put total casualties of Soviet Union's citizens alone to about 26 million.

Since 1945, the most commonly cited figure for the total number of Jews killed has been six million. The Yad Vashem Holocaust Martyrs' and Heroes' Remembrance Authority in Jerusalem, writes that there is no precise figure for the number of Jews killed. The figure most commonly used is the six million attributed to Adolf Eichmann, a senior SS official. Early calculations range from 5.1 million from Raul Hilberg, to 5.95 million from Jacob Leschinsky. Yisrael Gutman and Robert Rozett in the *Encyclopedia of the Holocaust* estimate 5.59-5.86 million. A study led by Wolfgang Benz of the Technical University of Berlin suggests 5.29-6.2 million. Yad Vashem writes that the main sources for these statistics are comparisons of prewar and postwar censuses and population estimates, and Nazi documentation on deportations and murders. Its Central Database of Shoah Victims' Names currently holds close to 3 million names of Holocaust victims, all accessible online. Yad Vashem continues its project of collecting names of Jewish victims from historical documents and individual memories.

Jews

Entrance to Auschwitz-Birkenau, 1945

Hilberg's estimate of 5.1 million, in the third edition of *The Destruction of the European Jews*, includes over 800,000 who died from "ghettoization and general privation"; 1,400,000 killed in open-air shootings; and up to 2,900,000 who perished in camps. Hilberg estimates the death toll of Jews in Poland as up to 3,000,000. Hilberg's numbers are generally considered to be a conservative estimate, as they typically include only those deaths for which records are available, avoiding statistical adjustment.

British historian Martin Gilbert used a similar approach in his *Atlas of the Holocaust*, but arrived at a number of 5.75 million Jewish victims, since he estimated higher numbers of Jews killed in Russia and other locations. Lucy S. Dawidowicz used pre-war census figures to estimate that 5.934 million Jews died.

There were about 8 to 10 million Jews in the territories controlled directly or indirectly by the Nazis (the uncertainty arises from the lack of knowledge about how many Jews there were in the Soviet Union). The six million killed in the Holocaust thus represent 60 to 75 percent of these Jews. Of Poland's 3.3 million Jews, over 90 percent were killed. The same proportion was killed in Latvia and Lithuania, but most of Estonia's Jews were evacuated in time. Of the 750,000 Jews in Germany and Austria in 1933, only about a quarter survived. Although many German Jews emigrated before 1939, the majority of these fled to Czechoslovakia, France or the Netherlands, from where they were later deported to their deaths. In Czechoslovakia, Greece, the Netherlands, and Yugoslavia, over 70 percent were killed. More than 50 percent were killed in Belgium, Hungary, and Romania. It is likely that a similar proportion was killed in Belarus and Ukraine, but these figures are less certain. Countries with notably lower proportions of deaths include Bulgaria, Denmark, France, Italy, and Norway. Albania was the only country occupied by the Nazis that had a significantly larger Jewish population in 1945 than in 1939. About two hundred native Jews and over a thousand refugees were provided with false documents, hidden when necessary, and generally treated as honored guests in a country whose population was roughly 60% Muslim.

Year	Jews killed
1933-1940	under 100,000
1941	1,100,000
1942	2,700,000
1943	500,000
1944	600,000
1945	100,000

The number of people killed at the major extermination camps is estimated as: Auschwitz-Birkenau: 1.4 million;[110] Treblinka: 870,000; Belzec: 600,000;[113] Majdanek: 79,000-235,000;[119][214] Chełmno: 320,000; Sobibór: 250,000. This gives a total of over 3.8 million; of these, 80-90% were estimated to be Jews. These seven camps thus accounted for half the total number of Jews killed in the entire Nazi Holocaust. Virtually the entire Jewish population of Poland died in these camps.

In addition to those who died in the above extermination camps, at least half a million Jews died in other camps, including the major concentration camps in Germany. These were not extermination camps, but had large numbers of Jewish prisoners at various times, particularly in the last year of the war as the Nazis withdrew from Poland. About a million people died in these camps, and although the proportion of Jews is not known with certainty, it was estimated to be at least 50 percent. Another 800,000 to one million Jews were killed by the *Einsatzgruppen* in the occupied Soviet territories (an approximate figure, since the *Einsatzgruppen* killings were frequently undocumented). Many more died through execution or of disease and malnutrition in the ghettos of Poland before they could be deported.

As most of the victims of the Holocaust were speakers of Yiddish, the Holocaust had a profound and permanent effect on the fate of Yiddish language and culture. On the eve of World War II, there were 11 to 13 million Yiddish speakers in the world. The Holocaust, however, led to a dramatic, sudden decline in the use of Yiddish, as the extensive Jewish communities, both secular and religious, that used Yiddish in their day-to-day life were largely destroyed. Around 5 million, or 85%, of the victims of the Holocaust, were speakers of Yiddish.

Non Jewish victims

One of Hitler's ambitions at the start of the war was to exterminate, expel, or enslave most or all Slavs from their native lands so as to make living space for German settlers. This plan of genocide was to be carried into effect gradually over a period of 25-30 years.

Ethnic Poles

Execution of Poles by *Einsatzkommando*,
Leszno, October 1939

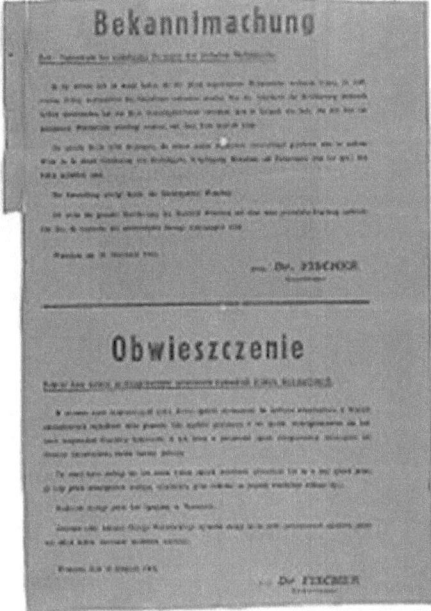

Announcement of death penalty for Poles helping Jews

Polish civilians executed in Warsaw

Auschwitz I patch with the letter "P", required wear for Polish inmates

Further information: Nazi crimes against ethnic Poles, Occupation of Poland (1939-1945), and Pacification operations in German-occupied Poland

German Nazi planners had in November 1939 called for "the complete destruction" of all Poles. "All Poles", Heinrich Himmler swore, "will disappear from the world". The Polish state under German occupation was to be cleared of ethnic Poles and settled by German colonists. Of the Poles, by 1952 only about 3-4 million of them were to be left in the former Poland, and only to serve as slaves for German settlers. They were to be forbidden to marry, the existing ban on any medical help to Poles in Germany would be extended, and eventually Poles would cease to exist. On August 22, 1939, about one week before the onset of the war, Hitler "prepared, for the moment only in the East, my 'Death's Head' formations with orders to kill without pity or mercy all men, women and children of Polish descent or language. Only in this way can we obtain the living space we need." Nazi planners decided against a genocide of ethnic Poles on the same scale as against ethnic Jews, it could not proceed in the short run since "such a solution to the Polish question would represent a burden to the German people into the distant future, and everywhere rob us of all understanding, not least in that neighbouring peoples would have to reckon at some appropriate time, with a similar fate".

The actions taken against ethnic Poles were not on the scale of the genocide of the Jews. Most Polish Jews (perhaps 90% of their antebellum population) perished during the Holocaust, while most Christian Poles (94%) survived the brutal German occupation. Between 1.8 and 2.1 million non-Jewish Polish citizens perished in German hands during the course of the war, about four-fifths of whom were ethnic Poles with the remaining fifth being ethnic minorities of Ukrainians and Belarusians, the vast majority of them civilians. At least 200,000 of these victims died in concentration camps with about 146,000 being killed in Auschwitz. Many others died as a result of general massacres such as in the Warsaw Uprising where between 120,000 and 200,000 civilians were killed. The policy of the Germans in Poland included diminishing food rations, conscious lowering of the state of hygiene and depriving the population of medical services. The general mortality rate rose from 13 to 18 per thousand. Overall, about 5.6 million of the victims WW2 were Polish citizens, both Jewish and non-Jewish, and over the course of the war Poland lost 16 percent of its pre-war population; approximately 3.1 million of the 3.3 million Polish Jews and approximately 2 million of the 31.7 million non-Jewish Polish citizens died at German hands during the war.[226] Over 90 percent of the death toll came through non-military losses, as most of the civilians were targeted by various deliberate actions by Nazi Germany and the Soviet Union.

Ethnic Serbs and South Slavs

In the Balkans, up to 581,000 Yugoslavs were killed by the Nazis and their Ustaše fascist allies in Yugoslavia.[227][228] German forces, under express orders from Hitler, fought with a special vengeance against the Serbs, who were considered Untermensch. The Ustaše collaborators conducted a systematic extermination of large numbers of people for political, religious or racial reasons. The most numerous victims were Serbs.

Bosniaks and Croats were also victims of Jasenovac. According to the U.S. Holocaust Museum:

"The Ustaša authorities established numerous concentration camps in Croatia between 1941 and 1945. These camps were used to isolate and murder Serbs, Jews, Roma, Muslims [Bosniaks], and other non-Catholic minorities, as well as Croatian political and religious opponents of the regime."

The USHMM and Jewish Virtual Library report between 56,000 and 97,000 persons were killed at the Jasenovac concentration camp. Yad Vashem reports an overall number of over 500,000 murders of Serbs "in horribly sadistic ways" at the hands of the Ustaša.

As per the most recent study, *Bosnjaci u Jasenovackom logoru* ("Bosniaks in Jasenovac concentration camp") by the author Nihad Halilbegovic, at least 103,000 Bosniaks (Bosnian Muslim Slavs) perished during Holocaust at the hands of the Nazi regime and Croatian Ustaše. According to the study "unknown is the full number of Bosniaks who were murdered under Serb or Croat alias or national name" and "large numbers of Bosniaks were killed and listed under Roma populations", therefore in advance sentenced to death and extermination.

East Slavs

In Belarus, Nazi Germany imposed a regime in the country that was responsible for burning down some 9,000 villages, deporting some 380,000 people for slave labour, and killing hundreds of thousands of civilians. More than 600 villages, like Khatyn, were burned along with their entire population and at least 5,295 Belarusian settlements were destroyed by the Nazis and some or all of their inhabitants killed. Altogether, 1,670,000 civilians (18 percent of the population) were killed during the three years of German occupation, including 245,000 Jews killed by the Einsatzgruppen.

Soviet POWs

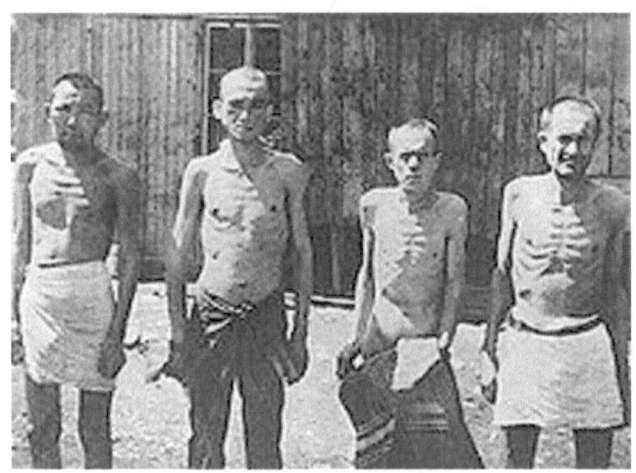

Soviet POWs in German captivity

According to Michael Berenbaum, between two and three million Soviet prisoners-of-war—or around 57 percent of all Soviet POWs—died of starvation, mistreatment, or executions between June 1941 and May 1945, and most those during their first year of captivity. According to other estimates by Daniel Goldhagen, an estimated 2.8 million Soviet POWs died in eight months in 1941-42, with a total of 3.5 million by mid-1944. The USHMM has estimated that 3.3 million of the 5.7 million Soviet POWs died in German custody—compared to 8,300 of 231,000 British and American prisoners.[238] The death rates decreased as the POWs were needed to work as slaves to help the German war effort; by 1943, half a million of them had been deployed as slave labor.

Romani people

Romani arrivals in the Belzec extermination camp, 1940

Map of persecution of the Roma

Because the Roma and Sinti are traditionally a secretive people with a culture based on oral history, less is known about their experience of the genocide than about that of any other group. Yehuda Bauer writes that the lack of information can be attributed to the Roma's distrust and suspicion, and to their humiliation, because some of the basic taboos of Romani culture regarding hygiene and sexual contact were violated at Auschwitz. Bauer writes that "most [Roma] could not relate their stories involving these tortures; as a result, most kept silent and thus increased the effects of the massive trauma they had undergone."

The treatment of Romanis was not consistent in the different areas that Nazi Germany conquered. In some areas (e.g. Luxembourg and the Baltic countries), the Nazis killed virtually the entire Romani population. In other areas (e.g. Denmark, Greece), there is no record of Romanis being subjected to mass killings.

Donald Niewyk and Frances Nicosia write that the death toll was at least 130,000 of the nearly one million Roma and Sinti in Nazi-controlled Europe. Michael Berenbaum writes that serious scholarly estimates lie between 90,000 and 220,000. A detailed study by the late Sybil Milton, formerly senior historian at the U.S. Holocaust Memorial Museum, calculated a death toll of at least 220,000 and possibly closer to 500,000, but this study explicitly excluded the Independent State of Croatia where the genocide of Romanies was intense.[244][245] Martin Gilbert estimates a total of more than 220,000 of the 700,000 Romani in Europe. Ian Hancock, Director of the Program of Romani Studies and the Romani Archives and Documentation Center at the University of Texas at Austin, has argued in favour of a higher figure of between 500,000 and 1,500,000. Hancock writes that, proportionately, the death toll equaled "and almost certainly exceed[ed], that of Jewish victims."

Before being sent to the camps, the victims were herded into ghettos, including several hundred into the Warsaw Ghetto. Further east, teams of Einsatzgruppen tracked down Romani encampments and murdered the inhabitants on the spot, leaving no records of the victims. They were also targeted by the puppet regimes that cooperated with the Nazis, e.g. the Ustaše regime in Croatia, where a large number of Romani were killed in the Jasenovac concentration camp. The genocide analyst Helen Fein has stated that the Ustashe killed virtually every Romani in Croatia.

In May 1942, the Romani were placed under the same labor and social laws as the Jews. On December 16, 1942, Heinrich Himmler, Commander of the SS and regarded as the "architect" of the Nazi genocide, issued a decree that "Gypsy *Mischlinge* (mixed breeds), Romani, and members of the clans of Balkan origins who are not of German blood" should be sent to Auschwitz, unless they had served in the Wehrmacht. On January 29, 1943, another decree ordered the deportation of all German Romani to Auschwitz.

This was adjusted on November 15, 1943, when Himmler ordered that, in the occupied Soviet areas, "sedentary Gypsies and part-Gypsies (*Mischlinge*) are to be treated as citizens of the country. Nomadic Gypsies and part-Gypsies are to be placed on the same level as Jews and placed in concentration camps." Bauer argues that this adjustment reflected Nazi ideology that the Roma, originally an Aryan population, had been "spoiled" by non-Romani blood.

Disabled and mentally ill

"60,000 RM is what this person with genetic defects costs the community during his lifetime. Fellow German, that's your money too"

Action T4 was a program established in 1939 to maintain the genetic purity of the German population by killing or sterilizing German and Austrian citizens who were judged to be disabled or suffering from mental disorder.

Between 1939 and 1941, 80,000 to 100,000 mentally ill adults in institutions were killed; 5,000 children in institutions; and 1,000 Jews in institutions. Outside the mental health institutions, the figures are estimated as 20,000 (according to Dr. Georg Renno, the deputy director of Schloss Hartheim, one of the euthanasia centers) or 400,000 (according to Frank Zeireis, the commandant of Mauthausen concentration camp). Another 300,000 were forcibly sterilized.[260] Overall it has been estimated that over 200,000 individuals with mental disorders of all kinds were put to death, although their mass murder has received relatively little historical attention. Despite not being formally ordered to take part, psychiatrists and psychiatric institutions were at the center of justifying, planning and carrying out the atrocities at every stage, and "constituted the connection" to the later annihilation of Jews and other "undesirables" in the Holocaust. After strong protests by the German Catholic and Protestant churches on August 24, 1941 Hitler ordered the cancellation of the T4 program.

The program was named after Tiergartenstraße 4, the address of a villa in the Berlin borough of Tiergarten, the headquarters of the *Gemeinnützige Stiftung für Heil und Anstaltspflege* (General Foundation for Welfare and Institutional Care), led by Philipp Bouhler, head of Hitler's private chancellery (*Kanzlei des Führer der NSDAP*) and Karl Brandt, Hitler's personal physician.

Brandt was tried in December 1946 at Nuremberg, along with 22 others, in a case known as *United States of America vs. Karl Brandt et al.*, also known as the Doctors' Trial. He was hanged at Landsberg Prison on June 2, 1948.

Homosexuals

The Homomonument in Amsterdam, a memorial to
the homosexual victims of Nazi Germany.

Between 5,000 and 15,000 homosexuals of German nationality are
estimated to have been sent to concentration camps. James D. Steakley
writes that what mattered in Germany was criminal intent or character,
rather than criminal acts, and the *"gesundes Volksempfinden"* ("healthy
sensibility of the people") became the leading normative legal principle.
In 1936, Himmler created the "Reichszentrale zur Bekämpfung der
Homosexualität und Abtreibung" ("Reich Central Office for the Combating
of Homosexuality and Abortion"). Homosexuality was declared contrary
to "wholesome popular sentiment," and homosexuals were consequently
regarded as "defilers of German blood." The Gestapo raided gay bars,

tracked individuals using the address books of those they arrested, used the subscription lists of gay magazines to find others, and encouraged people to report suspected homosexual behavior and to scrutinize the behavior of their neighbors.

Tens of thousands were convicted between 1933 and 1944 and sent to camps for "rehabilitation", where they were identified by yellow armbands and later pink triangles worn on the left side of the jacket and the right trouser leg, which singled them out for sexual abuse. Hundreds were castrated by court order They were humiliated, tortured, used in hormone experiments conducted by SS doctors, and killed. Steakley writes that the full extent of gay suffering was slow to emerge after the war. Many victims kept their stories to themselves because homosexuality remained criminalized in postwar Germany. Around two percent of German homosexuals were persecuted by Nazis.

The political left

German communists, socialists and trade unionists were among the earliest domestic opponents of Nazism and were also among the first to be sent to concentration camps. Hitler claimed that communism was a Jewish ideology which the Nazis termed "Judeo-Bolshevism". Fear of communist agitation was used as justification for the Enabling Act of 1933, the law which gave Hitler his original dictatorial powers. Hermann Göring later testified at the Nuremberg Trials that the Nazis' willingness to repress German communists prompted President Paul von Hindenburg and the German elite to cooperate with the Nazis. The first concentration camp was built at Dachau, in March 1933, to imprison German communists, socialists, trade unionists and others opposed to the Nazis. Communists, social democrats and other political prisoners were forced to wear a red triangle.

Hitler and the Nazis also hated German leftists because of their resistance to the party's racism. Many leaders of German leftist groups were Jews, and Jews were especially prominent among the leaders of the Spartacist uprising in 1919. Hitler already referred to Marxism and "Bolshevism" as a means of "the international Jew" to undermine "racial purity" and survival of the Nordics or Aryans, as well to stir up socioeconomic class tension and labor unions against the government or state-owned businesses. Within the concentration camps such as Buchenwald, German communists were privileged in comparison to Jews because of their "racial purity".

Whenever the Nazis occupied a new territory, members of communist, socialist, or anarchist groups were normally to be the first persons detained or executed. Evidence of this is found in Hitler's infamous Commissar Order, in which he ordered the summary execution of all political commissars captured among Soviet soldiers, as well as the execution of all Communist Party members in German held territory. Einsatzgruppen carried out these executions in the east.

Nacht und Nebel (German for "Night and Fog") was a directive (German: *Erlass*) of Hitler on December 7, 1941 signed and implemented by Chief of Staff of the Armed Forces Wilhelm Keitel, resulting in kidnapping and disappearance of many political activists throughout Nazi Germany's occupied territories.

Free masons

A memorial for *Loge Liberté chérie*, founded in November 1943 in Hut 6 of Emslandlager VII (KZ Esterwegen), one of two Masonic Lodges founded in a Nazi concentration camp.

In *Mein Kampf*, Hitler wrote that Freemasonry had "succumbed" to the Jews: "The general pacifistic paralysis of the national instinct of self-preservation begun by Freemasonry is then transmitted to the masses of society by the Jewish press." Freemasons were sent to concentration camps as political prisoners, and forced to wear an inverted *red triangle*. The United States Holocaust Memorial Museum believes "because many of the Freemasons who were arrested were also Jews and/or members of the political opposition, it is not known how many individuals were placed in Nazi concentration camps and/or were targeted only because they were Freemasons." However, the Grand Lodge of Scotland estimates the number of Freemasons executed between 80,000 and 200,000.

Jehovah's Witnesses

Refusing to pledge allegiance to the Nazi party or to serve in the military, roughly 12,000 Jehovah's Witnesses were forced to wear a purple triangle and were placed in camps where they were given the option of renouncing their faith and submitting to the state's authority. Between 2,500 and 5,000 were killed. Historian Detlef Garbe, director at the Neuengamme (Hamburg) Memorial, writes that "no other religious movement resisted the pressure to conform to National Socialism with comparable unanimity and steadfastness."

3—Expansion of the Universe

One of astronomical miracle of the Holy Quran is telling us that Allah built the sky and makes it more widely always. This has been mentioned in Surah Al-Zareiat-47.

Georges Lemaître proposed what became known as the Big Bang theory of the origin of the universe, although he called it his "hypothesis of the primeval atom". The framework for the model relies on Albert Einstein's general relativity and on simplifying assumptions (such as homogeneity and isotropy of space). The governing equations had been formulated by Alexander Friedmann. After Edwin Hubble discovered in 1929 that the distances to far away galaxies were generally proportional to their redshifts, as suggested by Lemaître in 1927, this observation was taken to indicate that all very distant galaxies and clusters have an apparent velocity directly away from our reference point: the farther away from the earth, the higher the apparent velocity If the distance between galaxy clusters is increasing today, This theory has been mentioned in Holy Quran who has told us since 1431 that Allah built up the sky and makes it more wide. This has been mentioned in Surah Al-zareiat-47. Everything must have been closer together in the past. This idea has been considered in detail back in time to extreme densities and temperatures, and large particle accelerators have been built to experiment on and test such conditions, resulting in significant confirmation of the theory, but these accelerators have limited capabilities to probe into such high energy regimes. Without any evidence associated with the earliest instant of the expansion, the Big Bang theory *cannot* and *does not* provide any explanation for such an initial condition; rather, it *describes* and *explains* the general evolution of the universe since that instant. The observed abundances of the light elements throughout the cosmos closely match the calculated predictions for the formation of these elements from nuclear processes in the rapidly expanding and cooling first minutes of the universe, as logically and quantitatively detailed according to Big Bang nucleo-synthesis.

Of course, this information of the expansion of the universe was not available absolutely since 1431 years between people dwelling the desert of Mecca in KSA.

4—Black Holes

In Holy Quran, in Surah Al-Takweer-15, 16 Allah swears with the running invisibles moving to swallow (i.e. attract asteroids).

According to the general theory of relativity, a **black hole** is a region of space from which nothing, not even light, can escape. It is the result of the deformation of space-time caused by a very dense compact mass. Around a black hole there is an undetectable surface which marks the point of no return, called an event horizon. It is called "black" because it absorbs all the light that hits it, reflecting nothing, just like a perfect black body in thermodynamics. Under the theory of quantum mechanics, black holes possess a temperature and emit Hawking radiation, but for black holes of stellar mass or larger this temperature is much lower than that of the cosmic background radiation.

Despite its invisible interior, a black hole can be observed through its interaction with other matter. A black hole can be inferred by tracking the movement of a group of stars that orbit a region in space. Alternatively, when gas falls into a stellar black hole from a companion star, the gas spirals inward, heating to very high temperatures and emitting large amounts of radiation that can be detected from earthbound and Earth-orbiting telescopes.

Astronomers have identified numerous stellar black hole candidates, and have also found evidence of super-massive black holes at the center of galaxies. In 1998, astronomers found compelling evidence that a super-massive black hole of more than 2 million solar masses is located near the Sagittarius A* region in the center of the Milky Way galaxy, and more recent results using additional data find evidence that the super-massive black hole is more than 4 million solar masses.

I don't like to be unrestrained in the future events in the Holy Quran because it has many future events and past events in addition to events in the other universe (Paradise and hell). Also Holy Quran has astronomical miracles.

From the previous discussion we can abbreviate the conditions of the comforter who tells all the truth from Allah, God of all, as follows:

1—He is Illiterate 2—He is Shepherd 3—He tells us what he heard 4—He tells us with the future events 5—He verifies the message of Jesus as a prophet from Allah via the Holy Quran by the Holy Spirit. 6-He has the light which had been descended onto him by the Holy Sprit, Gabriel, from Allah.

Finally we can say:

1— Allah puts his name in the right hand of man in Arabic language

2— The country of Arabic language after Jesus by about five centuries is Hegaz which is now Kingdom of Saudi Arabia.

3— The birth place of the Arabic prophet is in Mecca.

4— Allah creates Man in the skeleton of Mohammed in Arabic language.

5— Mohammed was illiterate and shepherd.

6— The Holy Quran had been descended on him and the Holy Quran is the light from the heaven.

Finally Allah with his hand possesses every thing created by him in the universe Space from atoms to galaxies. So Allah is The God of everything.

References

1— From Wikipedia, the free encyclopedia

2— **Holy Quran**

3— **Discovery (New Testament) National Student Ministries by American Bible Society 1966, 1971.**

4— **The Holy Bible, Living Edition, (Poket Edition), Tyndale House Publisher (1989).**

5— **الكتاب المقدس (كتاب الحياة-الترجمة التفسيرية)**
(1988)
ISBN 086660 407 3 Blue
ISBN 086660 408 1 Red
ISBN 086660 409 X Brown

www.ingramcontent.com/pod-product-compliance
Lightning Source LLC
Chambersburg PA
CBHW031838170526
45157CB00001B/349